SpringerBriefs in Applied Sciences and Technology

Computational Intelligence

Series Editor

Janusz Kacprzyk, Systems Research Institute, Polish Academy of Sciences,
Warsaw, Poland

SpringerBriefs in Computational Intelligence are a series of slim high-quality publications encompassing the entire spectrum of Computational Intelligence. Featuring compact volumes of 50 to 125 pages (approximately 20,000-45,000 words), Briefs are shorter than a conventional book but longer than a journal article. Thus Briefs serve as timely, concise tools for students, researchers, and professionals.

Fivos Papadimitriou

Spatial Artificial Intelligence

 Springer

Fivos Papadimitriou
Mathematisch-Naturwissenschaftliche
Fakultät
University of Tübingen
Tübingen, Germany

ISSN 2191-530X   ISSN 2191-5318 (electronic)
SpringerBriefs in Applied Sciences and Technology
ISSN 2625-3704   ISSN 2625-3712 (electronic)
SpringerBriefs in Computational Intelligence
ISBN 978-3-031-82135-6   ISBN 978-3-031-82136-3 (eBook)
https://doi.org/10.1007/978-3-031-82136-3

This Springer imprint is published by the registered company Springer Nature Switzerland AG
The registered company address is: Gewerbestrasse 11, 6330 Cham, Switzerland

If disposing of this product, please recycle the paper.

Dedicated to Nionioki

Preface

Spatial data (maps, images, solid objects, etc.) encapsulate two or more spatial dimensions and are used in multiple geographic, environmental, social, medical, engineering and other applications. A large part of them is suitable for AI-based analyses, so they are increasingly processed with the aid of AI and, quite probably, the attractiveness of AI nowadays is partly due to the newly invented and remarkable AI methods that enable us to process, analyze, manipulate and synthesize spatial data in ways previously unimagined of. As a matter of fact, Spatial AI is already well-developed, and its potential is being intensively exploited by scientists and engineers around the world.

This book aims to address the key concepts, basic methods, most common applications and possible futures of this kind of AI.

Spatial AI is defined here as the AI which is

– used *for spatial* analysis and spatial problem-solving;
– generated *from spatial* data;
– embedded *in spatial* domains (physical and/or digital),
– (or any combinations of the above).

Chapter 1 examines the theoretical underpinnings and the contextual framework of Spatial AI, as documented from qualitative and quantitative analyses of the relevant scientific literature.

Chapter 2 shows how AI has been used *for* spatial analysis, by presenting the first stage of development of Spatial AI that was mainly based on Symbolic AI and Evolutionary AI.

Chapter 3 exposes the methods by which AI can be generated *from* spatial data and traces the next stage of development of Spatial AI, through Neurosymbolic to Generative AI, by using LLMs, CNNs, DNNs, etc. and, eventually, ChatGPT.

Chapter 4 focuses on AI embedded *in* digital and/or physical spatial settings. The "spatialization" of AI consists in the deployment of Spatial AI code or software in Robotics, Ambient Intelligence or Spatial Computing (involving Augmented Reality, the Metaverse and Digital Twins).

Chapter 5 defines specific criteria that might be useful in defining spatially-enabled Artificial General Intelligence (AGI), by appraising specific characteristics and capabilities that AGI should have in order to think and act, in spatial terms, like humans.

Chapter 6 explores the key properties that Artificial SuperIntelligence (ASI) should be endowed with in order to surpass both AGI and humans in its capacity to identify spatial entities, to process spatial data and solve spatial problems.

Chapter 7 examines the impact of possible physical limitations to the growth of Spatial AI that might lead to transcomputation, while also taking into account the prospects that Quantum AI sets out for the future of spatial computation.

Chapter 8 invites the reader to reflect on the ultimate mathematical and computational limitations to the capabilities of Spatial AI that are imposed by spatial complexity and/or computational complexity.

Spatial AI is in a phase of rapid expansion and there is every reason to believe that this expansion will continue, with even more impressive results in the years to come. Plausibly, certain philosophical, ethical or legal problems will emerge due to its growth, but these are beyond the scope of this book.

So what is written here is in the hope that all what *we, humans*, instruct Spatial AI to deliver to *us*, will ultimately serve *us* only, for *our own* prosperity and happiness, as well as for improving the living conditions of as many sentient beings as possible.

Tübingen, Germany   Fivos Papadimitriou

Contents

Acronyms

ABM	Agent-Based Models
ACO	Ant Colony Optimization
AGI	Artificial General Intelligence
AHI	Artificial Human Intelligence
AI	Artificial Intelligence
ALife	Artificial Life
AR	Augmented Reality
ASI	Artificial Superintelligence
CNN	Convolutional Neural Network
DNN	Deep Neural Network
GAN	Generative Adversarial Network
GeoAI	Geospatial Artificial Intelligence
GPT	Generative Pre-trained Transformer
LLM	Large Language Model
NLP	Natural Language Processing
NN	Neural Network
QAI	Quantum Artificial Intelligence
SLAM	Simultaneous Localization and Mapping
SVI	Search Volume Indices
SVM	Support Vector Machines
UAV	Unmanned Aerial Vehicles
VR	Virtual Reality
XNN	Explainable Neural Network

Chapter 1
The Rise of Spatial AI

Abstract Spatial AI is defined here as AI *for* spatial analysis, AI that is generated/ created *from* spatial data and AI that is embedded *in* physical and/or digital spatial domains (or any combinations of these). Spatial AI entails Geospatial AI (GeoAI), as well as several forms of spatially-enabled AI. Some characteristic applications of Spatial AI are considered in this chapter, in various fields of science and technology, along with scientometric analyses of the key concepts and numbers of publications that define the context and boundaries of this rapidly developing field of AI. An examination of the conceptual framework of Spatial AI in comparison with GeoAI shows that although GeoAI has emerged as the prevailing type of Spatial AI it is not the only one since other domains of AI also contribute to the development of theory and applications of Spatial AI. Indeed, these domains (spatial robotics, AR, metaverse, medical applications of Spatial AI, applications of evolutionary AI etc.) are spatial but not geographic.

Keywords Spatial AI · GeoAI · Scientometry · Convolutional neural networks · AI in medicine · Spatial robotics · Spatial computing · Evolutionary AI

1.1 Spatial AI and GeoAI

The onset of Geospatial AI (GeoAI) can be traced back to the AI-Geography relationships in the middle 1980s, with the works of Smith (1984) and Couclelis (1986). Having received a significant boost in the middle 1990s (Openshaw, 1992; Openshaw & Openshaw, 1997) with "geocomputation", the research that was carried out in AI ever since endowed geographical analysis with entirely new ideas, models and methods. Nowadays, Geography and AI seem to have jointly shaped a new vivid field of research and applications, GeoAI, which integrates concepts and methods of typical fields of AI such as machine learning, deep learning and artificial neural networks with geospatial analysis (Janowich et al., 2020). Parallel to Spatial AI, and akin to the usefulness of AI in modeling spatial dynamics for geographical applications, is the concept of "environmental intelligence", which consists in integrating diverse environmental data of a particular geographical region with the aim of sustainable management. Expectedly, such data are large and complex, derived i.e. from satellite observations, drones, field observations and photographs. Presently, there are no clear cut boundaries between environmental intelligence and GeoAI (see Chen et al., 2008), although the latter also aims at exploiting integrated spatial data sets in order to derive intelligent automated decisions (Bhadran et al., 2018; Bhatti, 2020; Döllner, 2020).

During the last years, AI has proven particularly effective in analyzing geographical changes from satellite imagery, particularly with the aid of convolutional neural networks (CNNs) which have revolutionized AI by endowing it with a distinctive spatial character. With the term CNN rapidly becoming widely known, it is important to notice the basic difference that these networks have in terms of structure compared to artificial neural networks: the key difference consists in the operation of "convolution", which is an unusual (but, as it turned out, utterly significant) form of numerical operation involving only additions and multiplications of the pixel values of images. Consequently, by using methods of deep learning (Zhu et al., 2017; Touya et al., 2019; Zhang et al., 2019), CNNs or even deep convolutional neural networks (DCNNs) were widely used to automatically classify land use/land cover (Huang et al., 2018; Law et al., 2020; Scott et al., 2017) and terrain features (Li & Hsu, 2020). Google's deep neural network model (DNN) "Inception" was launched in 2014 and was able to automatically classify images by performing 1.5 billion arithmetic operations and using 22 levels. Two years later, DNNs became able to outperform humans in image-identification tasks while using petabytes of satellite imagery and statistical data. The usefulness of CNNs and DNNs in the classification of satellite imagery thus contributed to the field of Geospatial AI, or GeoAI, which evolved as the main pillar of Spatial AI.

Surprisingly though, one of the applications of such networks, along with the further development of generative adversarial networks (GANs) has been the creation of fake satellite images presenting non-existent geographical features (Goodfellow et al., 2014) and so geospatial data can now be used to create fake maps and fake landscape imagery (Zhao & Sui, 2017; Zhang et al., 2019a, 2019b; Zhang et al., 2020; Zhao et al., 2021). The central idea behind AI-made fake maps is that it is possible to use the base map of a geographical region A, combine it with the satellite image of a region B and so produce a new satellite image of region A which has inherited a mixture of characteristics of both A and B (Zhao et al., 2021).

The rise of Spatial AI can be scientometrically documented by using the software "Vos Viewer" (Van Eck & Waltman, 2010) which surveys the scientific literature and then renders visualizations of the co-occurrence of concepts (keywords) represented as nodes with the thicknesses of links among nodes being proportional to the number of publications that share the same pair of concepts as the pair of nodes that each link connects. Beneath each network appears a time bar displaying the mean publication times per concept (node) and per link. Datasets of the relevant scientific literature (i.e. Scopus, Openalex) can be analysed, using keywords (i.e. "Spatial AI", "GeoAI" etc.) and so create networks linking concepts with varying link strengths. The scientometric analysis of the concepts for each dataset of publications related to GeoAI for instance (Fig. 1.1) shows that three concepts (computer science, geography and AI) are by far the most important ones, whereas the number of publications on GeoAI presents a fast increase after 2010.

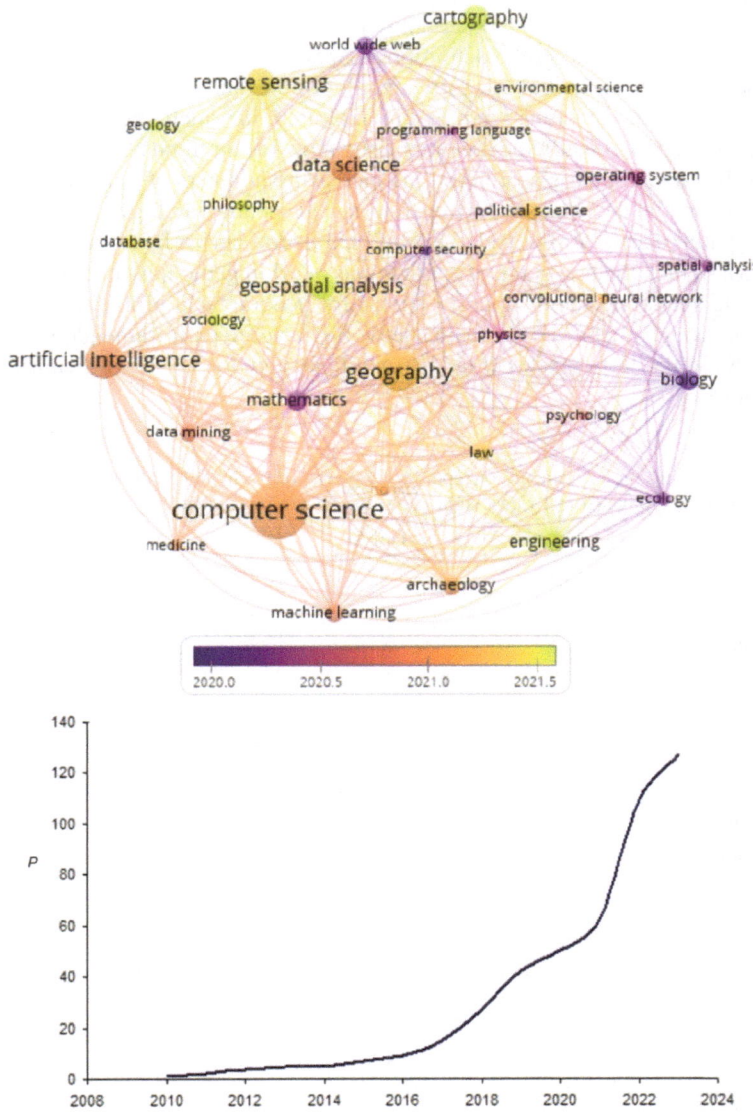

Fig. 1.1 The prominent concepts in the literature of GeoAI (above) and the mean annual number of publications (below)

Besides the literature on GeoAI, two more datasets of publications in the scientific literature are analyzed in comparison (Geospatial AI, GIS AI), with respect to the ten most important concepts that were ranked by decreasing order (Table 1.1) for each dataset. The pattern by which keywords are ranked in these four interrelated datasets (Spatial AI, GeoAI, Geospatial AI, GIS AI) reveals that the context of Spatial AI permeates all the other datasets (Fig. 1.2), while "Computer Science" and "Geography" prevail in the definition of Spatial AI.

In the literature of Spatial AI, software, algorithms and datasets for deep spatial learning and spatial machine learning are exploited, in which cartography, remote sensing, geology and medicine are among the most recent trends (Fig. 1.3).

A further scientometric search reveals increasing numbers of publications in the domains of spatial decision support systems and geospatial intelligence (Fig. 1.4), as well as spatial deep learning and convolutional neural networks for image analysis (Fig. 1.5). In fact, the first stage of applications of AI in geography mainly consisted in devising spatial decision support systems for geographical problem-solving, but the breakthroughs achieved in neural networks have given a tremendous boost to Spatial AI in the last decades (Fig. 1.6) while also endowing it with entirely new capabilities.

Table 1.1 Ranking the prominent concepts in the literature of Spatial AI, GeoAI Geospatial AI and GIS AI: ranks range from 1 (least significant) to 10 (most significant)

Concepts	Spatial AI	GeoAI	Geospatial AI	GIS AI
Computer science	10	10	9	10
AI	9	8	5	
Geography	8	9	8	9
Mathematics	7	4	2	2
Ecology	6			
Machine learning	5			
Geology	4			
Engineering	3		3	8
Data mining	2	3	1	3
Remote sensing	1	2	7	4
GIS		1		6
Geospatial analysis		7	10	
Data science		6	6	
Cartography		5	4	7
Art				1

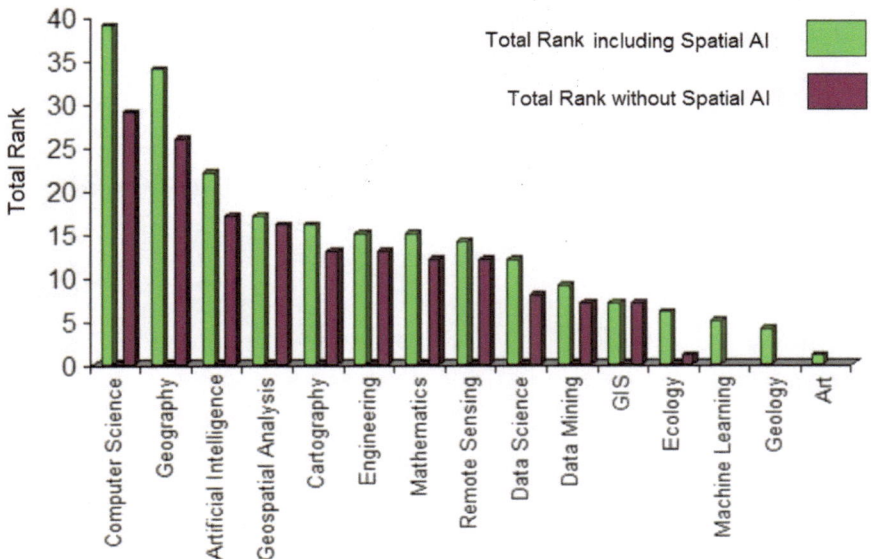

Fig. 1.2 Ranks of keywords for all four datasets in the scientific literature for Spatial AI, GeoAI, Geospatial AI and GIS AI

1.2 Spatial AI Beyond GeoAI

Despite the centrality of GeoAI in the range of topics that are considered as "spatial" in AI, the field of Spatial AI is much wider than GeoAI, because Spatial AI encompasses research methods and technological applications which are *spatial but not geographical.*

Robot mapping for instance is a domain of AI that is "spatial" but not "geographical"; a spatial AI-related domain of robotics and spatial robotics is also a domain of Spatial AI (Kuipers & Byun, 1991; Belfiore & Di Benedetto, 2000; Zender et al., 2008) and its literature presents a steady growth (Fig. 1.7), whereas SLAM (Simultaneous Localization and Mapping) robotics has emerged as the prominent research field in "spatial robotics" (Fig. 1.8).

Besides spatial robotics however, life-like entities in silica, artificial societies and artificial ecosystems that "live" in "programming environments" are visible on computer screens and have acquired spatial characteristics (i.e. "artificial life" entities). While cyber-physical systems are already used in a wide variety of applications (including ecological research), in the long run they may as well comprise diverse physical and information entities embedded in (or unified with) natural-and-artificial ecosystems, of which the components may interact, collaborate, adopt to their environment or to one another, even evolve (Moldovan et al., 2018) and, in so doing, exhibit (under certain conditions) "intelligent" behaviours.

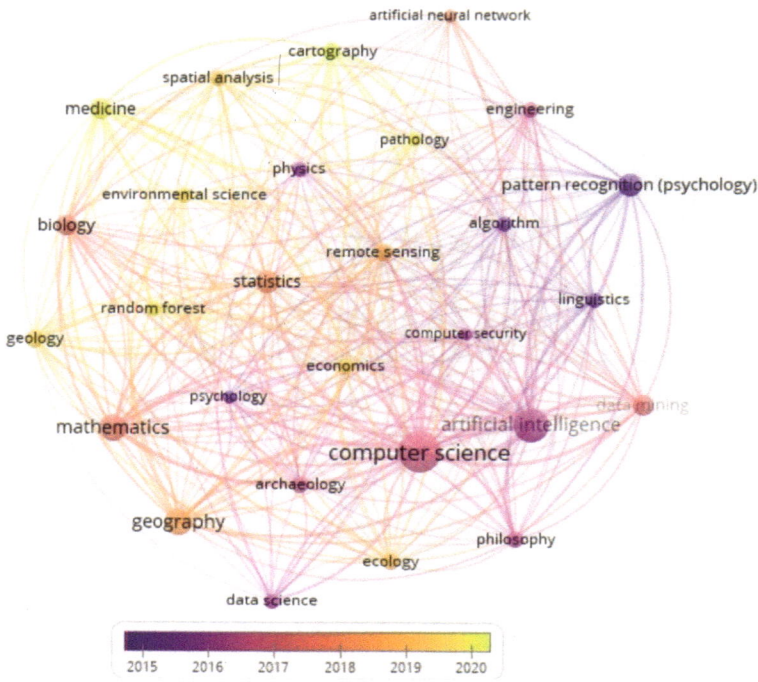

Fig. 1.3 The most important concepts in the literature of spatial deep learning and spatial machine learning

In the efforts to design, create and, moreover, simulate such hybrid ecosystems, concepts of ecology have repeatedly been used in bio-inspired computation, as well as in cyber-physical ecosystems and robotics, to the extent that we can even talk about "robotic ecologies" (Dragone et al., 2015). Further still, by mimicking the evolutionary processes of living organisms, a separate branch of AI has developed, the "Evolutionary AI", which exploits applications of genetic algorithms and swarm intelligence (Fig. 1.9), in order to simulate a great variety of forms of non-human intelligence (Papadimitriou, 2022a, 2022b, 2022c), whereas the scientific literature of Evolutionary AI displays a shift of emphasis towards medicine (and genetics), data mining and statistics.

Aside of Evolutionary AI (of which Swarm AI is part of), new ways of thinking through synergies with AI have opened up in the domain of Spatial Computing. Perhaps, one of the most impressive examples to be found is the intersection of augmented reality (AR) with AI (Devagiri et al., 2022), bringing the spatialization of AI to an altogether different state which involves artificial environments that give the sense of immersion (Fig. 1.10). Furthermore, spaces created with Augmented Reality (AR) are instrumental in building the Metaverse (Guo et al., 2022) and Digital Twins (notice that neither AR, nor the Metaverse, nor even Digital Twins

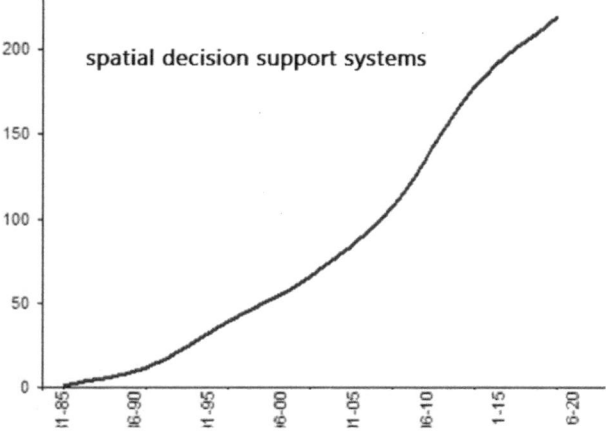

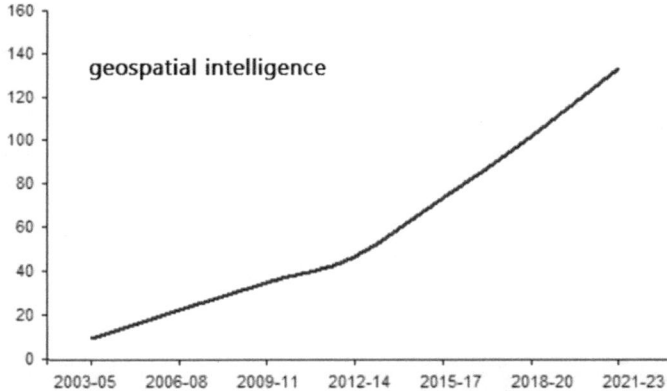

Fig. 1.4 The quantity of publications on spatial decision support systems and geospatial intelligence

are considered to be basic components of GeoAI). As the literature reveals, the term "Metaverse" is primarily associated to data science, knowledge management, law and computer security (Fig. 1.11), while, as the scientific literature reveals, geography, psychology and medicine are some of the newest trends in the field of intersections between AI and Digital Twins (Fig. 1.12).

But the scope of Spatial AI is even wider; it has applications in medicine also, as CNNs have proven useful in MRI analysis (Fig. 1.13) for the automatic derivation of medical diagnoses: brain tumors (Bernal et al., 2019; Gull & Akbar, 2021; Chattopadhyay & Maitra, 2022; Aleid et al., 2023), prostate cancer (Gunashekar et al., 2022), skeletal radiology (Fritz & Fritz, 2022), nasopharyngeal carcinomas (Wong et al., 2021), classification of aortic stenoses (Elvas et al., 2023), examination of

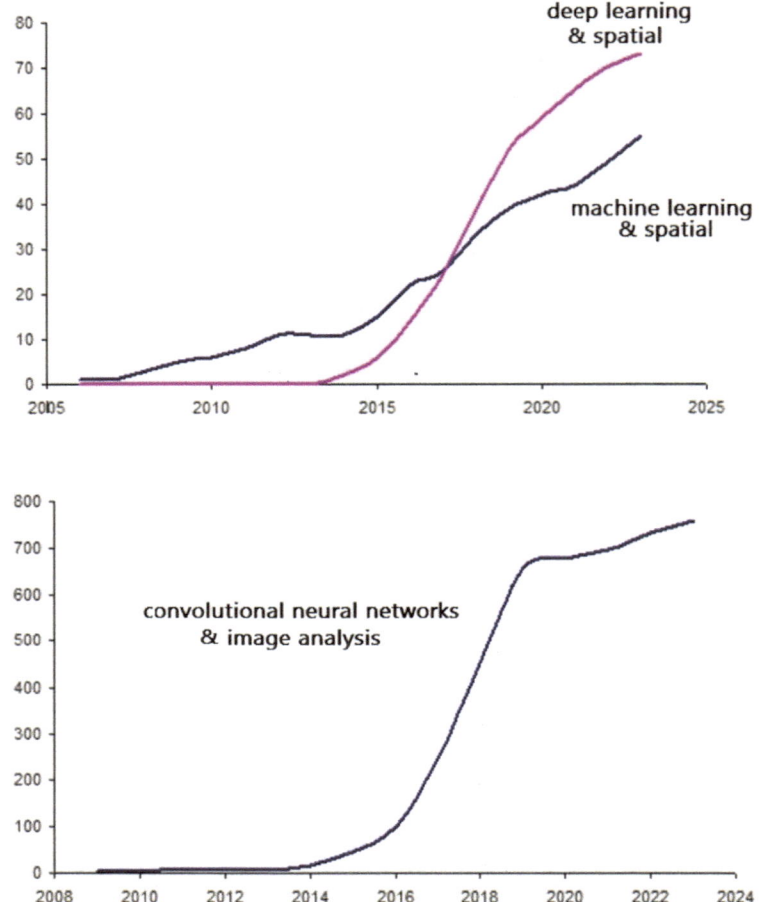

Fig. 1.5 The increasing numbers of publications on spatial deep learning, spatial machine learning and convolutional neural networks for image analysis

breast lesions (Lo Gullo et al., 2024), detecting Alzheimer disease (Bi et al., 2021; Jain et al., 2021) etc.

These considerations lead us to perceive Spatial AI as an emerging scientific domain of AI that is endowed with its own dynamics and related to (at least) five different scientific fields other than AI (Fig. 1.14).

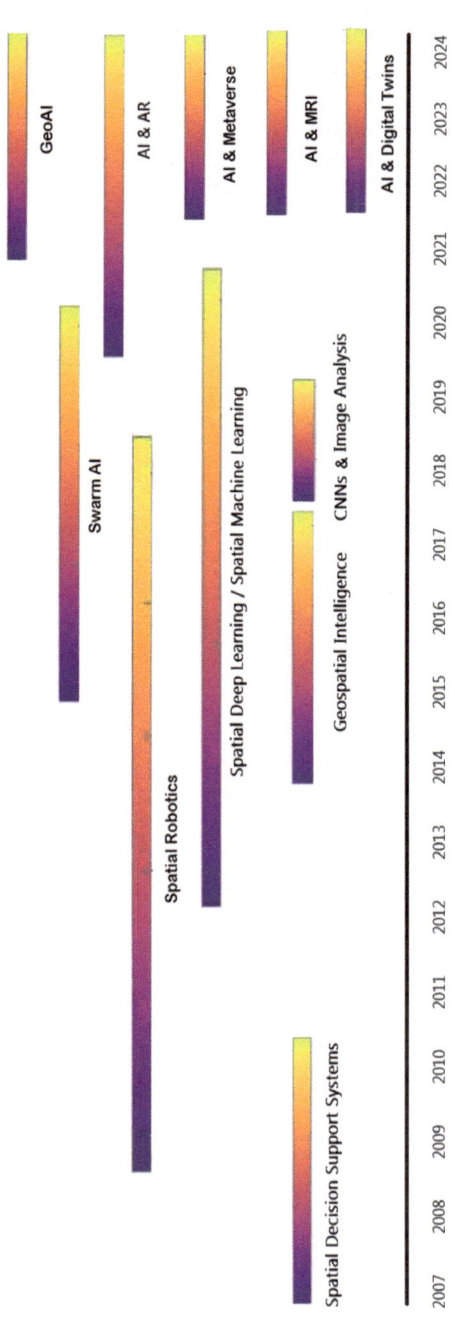

Fig. 1.6 The ranges of mean publication times of scientific publications related to Spatial AI

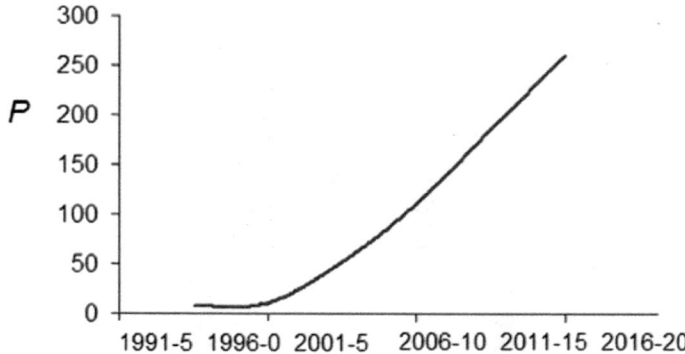

Fig. 1.7 The annual number of publications in the field of "robot mapping" (source of data: Openalex)

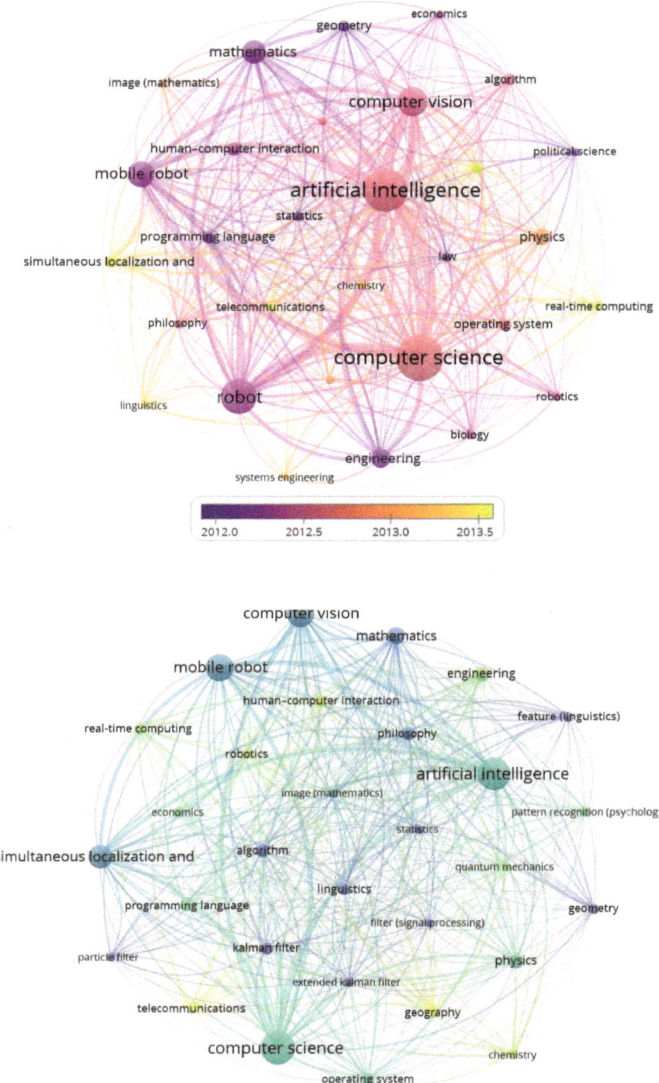

Fig. 1.8 The main concepts and publication times in the literature of spatial robotics (above) and SLAM (below)

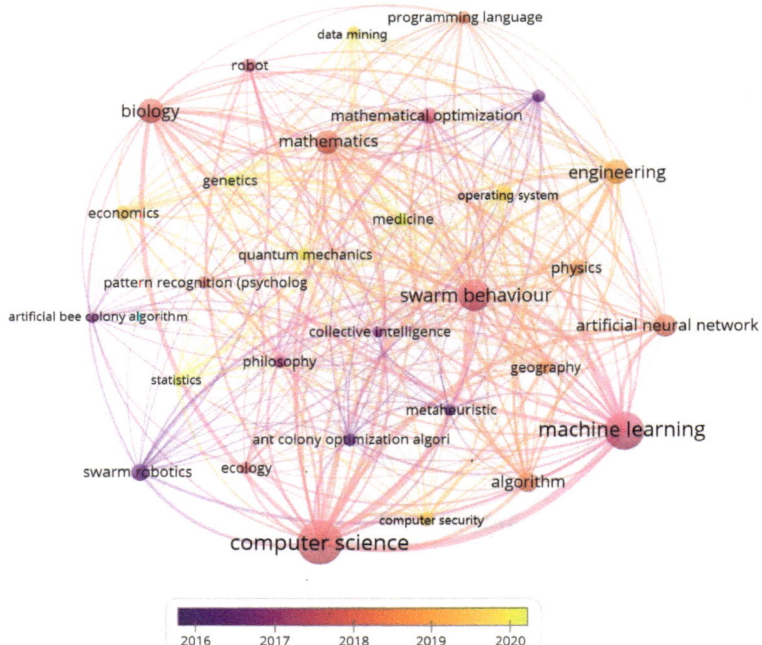

Fig. 1.9 The main concepts and mean publication times in the literature of Swarm Artificial Intelligence

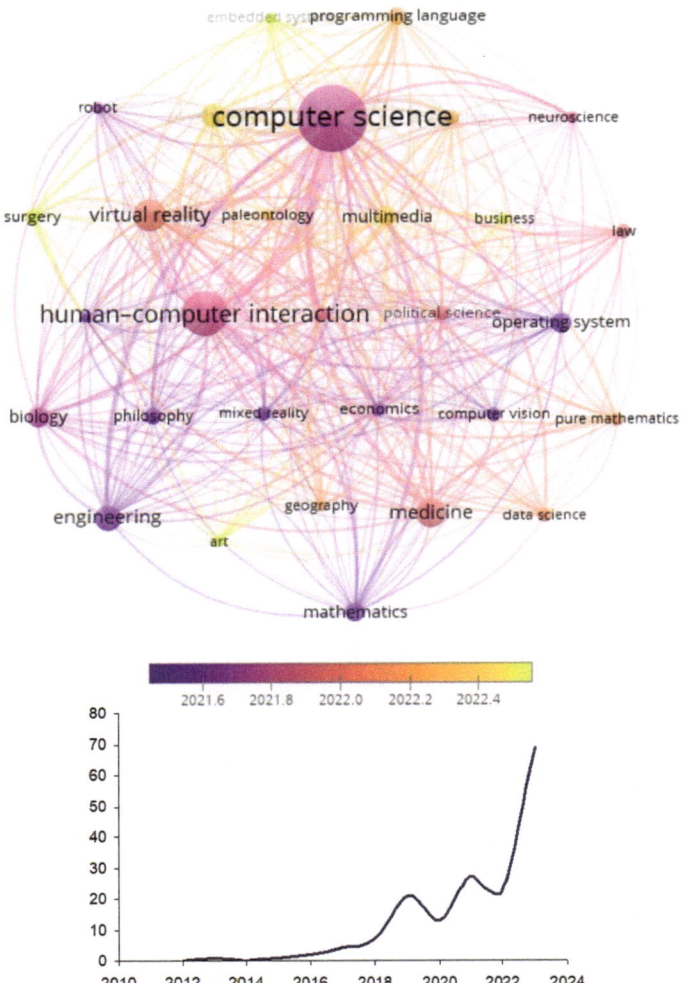

Fig. 1.10 The 30 most important keywords in the literature of AR & AI (above) and the increase in the number of papers in the last years (below)

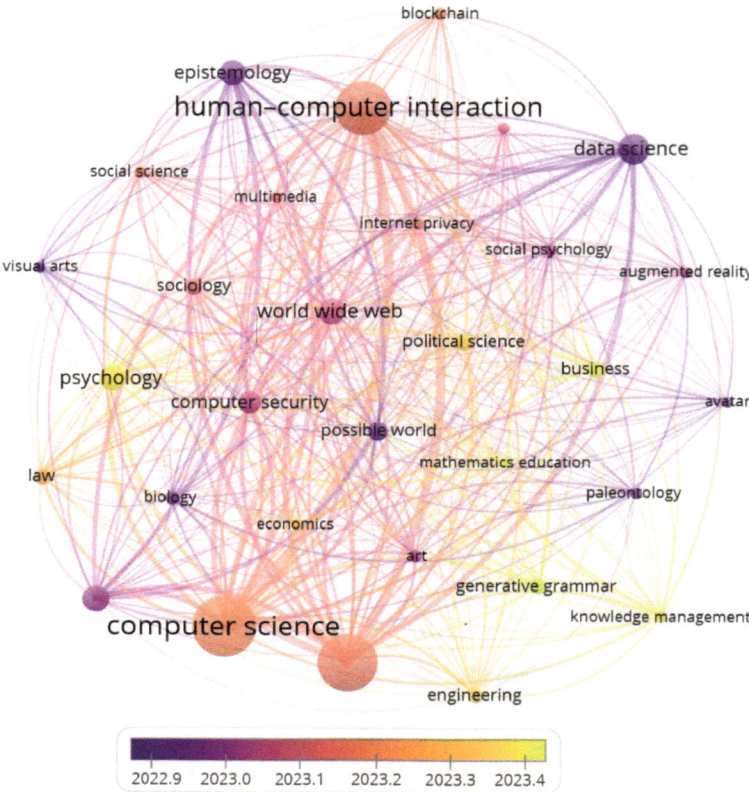

Fig. 1.11 The 30 most important keywords in the literature of AI & Metaverse

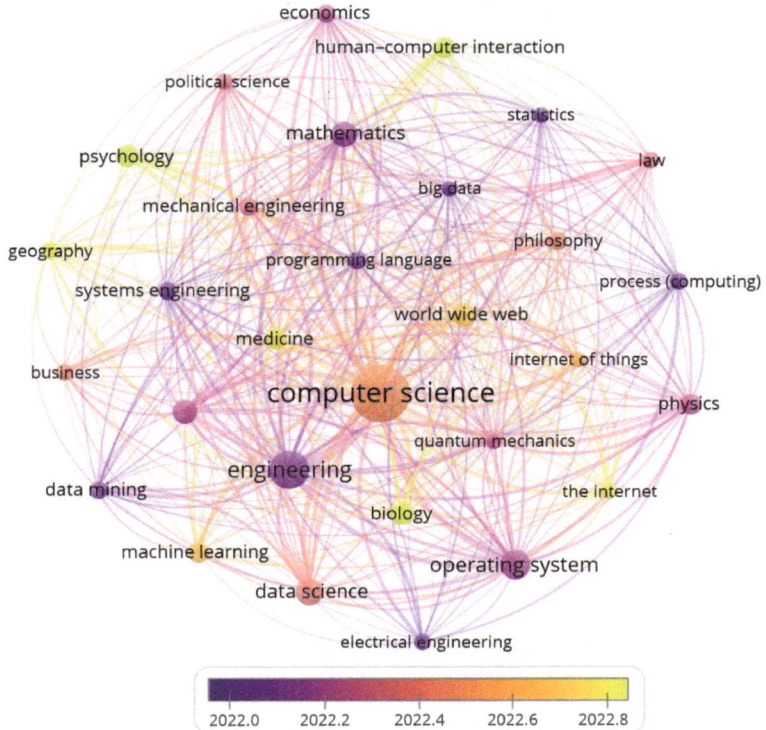

Fig. 1.12 The main concepts in the literature of AI applications in Digital Twins

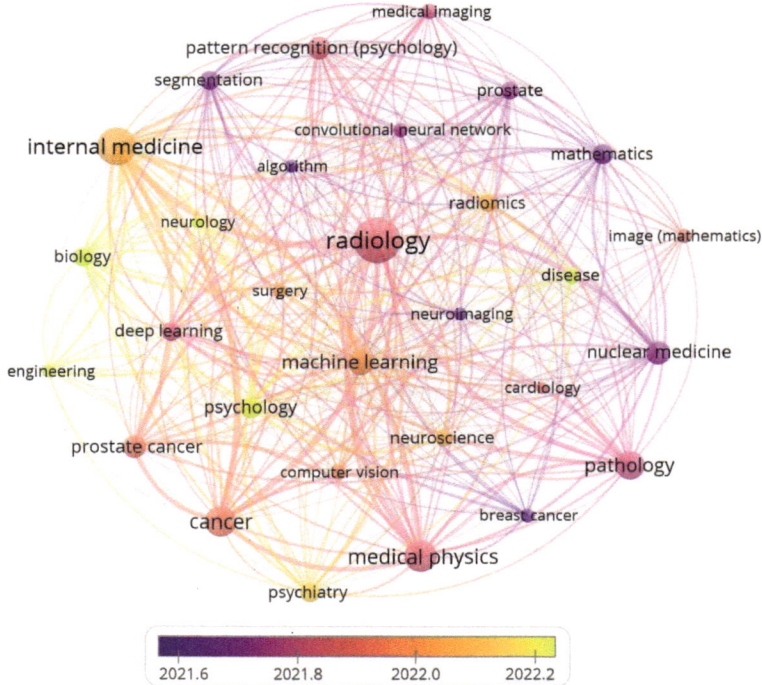

Fig. 1.13 The main concepts in the literature related to applications of Spatial AI in medicine: MRI analysis using AI methods

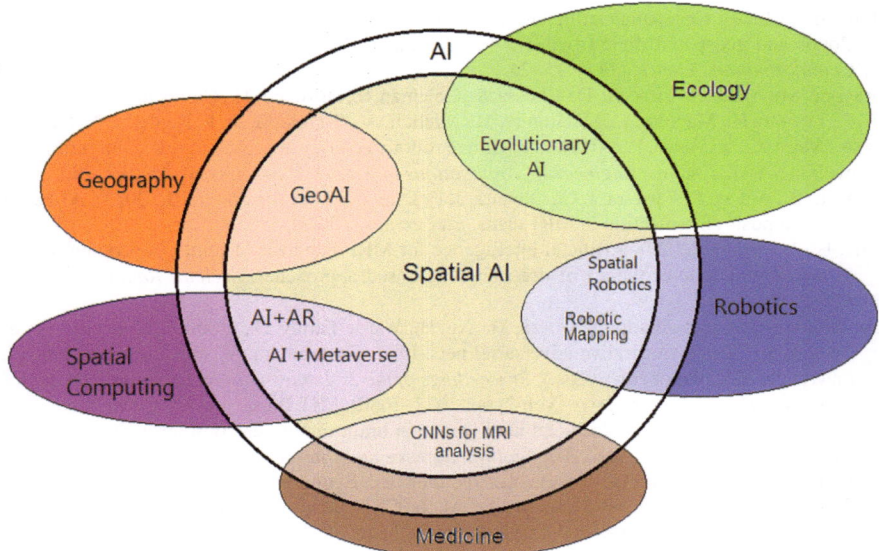

Fig. 1.14 Scientific fields that relate to AI and the intersections that define the field of Spatial AI

References

Aleid, A., Alhussaini, K., Alanazi, R., Altwaimi, M., Altwijri, O., & Saad, A. S. (2023). Artificial intelligence approach for early detection of brain tumors using MRI images. *Applied Sciences, 13*(6), 3808.

Belfiore, N. P., & Di Benedetto, A. (2000). Connectivity and redundancy in spatial robots. *The International Journal of Robotics Research, 19*(12), 1245–1261.

Bernal, J., Kushibar, K., Asfaw, D. S., Valverde, S., Oliver, A., Martí, R., & Lladó, X. (2019). Deep convolutional neural networks for brain image analysis on magnetic resonance imaging: A review. *Artificial Intelligence in Medicine, 95*, 64–81.

Bhadran, A., & Girishbai, D. (2018). Future road map for Geodata towards Geospatial Artificial Intelligence. *Indian Journal of Geosciences, 72*(2), 133–138.

Bhatti, U. A., Yu, Z., Yuan, L., Zeeshan, Z., Nawaz, S. A., Bhatti, M., Mehmood, A., Ain, Q. U., & Wen, L. (2020). Geometric algebra applications in geospatial artificial intelligence and remote sensing image processing. *IEEE Access, 8*, 155783–155796.

Bi, X., Liu, W., Liu, H., & Shang, Q. (2021). Artificial Intelligence-based MRI Images for Brain in Prediction of Alzheimer′s Disease. *Journal of Healthcare Engineering, 2021*(1), 8198552.

Chattopadhyay, A., & Maitra, M. (2022). MRI-based brain tumour image detection using CNN based deep learning method. *Neuroscience Informatics, 2*(4), 100060.

Chen, S. H., Jakeman, A. J., & Norton, J. P. (2008). Artificial intelligence techniques: An introduction to their use for modelling environmental systems. *Mathematics and Computers in Simulation, 78*(2–3), 379–400.

Couclelis, H. (1986). Artificial intelligence in geography: Conjectures on the shape of things to come. *Professional Geographer, 38*(1), 1–11.

Devagiri, J. S., Paheding, S., Niyaz, Q., Yang, X., & Smith, S. (2022). Augmented reality and artificial intelligence in industry: Trends, tools, and future challenges. *Expert Systems with Applications, 207*, 118002.

Döllner, J. (2020). Geospatial artificial intelligence: Potentials of machine learning for 3d point clouds and geospatial digital twins. *PFG–Journal of Photogrammetry, Remote Sensing and Geoinformation Science, 88(1)*, 15–24.

Dragone, M., Amato, G., Bacciu, D., Chessa, S., Coleman, S., DiRocco, M., Gallicchio, C., Gennaro, C., Lozano, H., Maguire, L., McGinnity, M., Micheli, A., O'Hare, G. M. P., Renteria, A., Saffioti, A., Vairo, C., & Vance, P. (2015). A cognitive robotic ecology approach to self-configuring and evolving AAL systems. *Engineering Applications of Artificial Intelligence, 45*, 269–280.

Elvas, L. B., Águas, P., Ferreira, J. C., Oliveira, J. P., Dias, M. S., & Rosário, L. B. (2023). AI-based aortic stenosis classification in MRI scans. *Electronics, 12(23)*, 4835.

Fritz, B., & Fritz, J. (2022). Artificial intelligence for MRI diagnosis of joints: A scoping review of the current state-of-the-art of deep learning-based approaches. *Skeletal Radiology, 51(2)*, 315–329.

Goodfellow, I., Pouget-Abadie, J., Mirza, M., Xu, B., Warde-Farley, D., Ozair, S., Courville, A., & Bengio, Y. (2014). Generative adversarial nets. In Z. Ghahramani, M. Welling, C. Cortes, N. Lawrence, & K. Weinberger (Eds.), *Proceedings of the 27th international conference on neural information processing systems* (Vol. 2, pp. 2672–2680). MIT Press.

Gull, S., & Akbar, S. (2021). Artificial intelligence in brain tumor detection through MRI scans: advancements and challenges. *Artificial Intelligence and Internet of Things*, 241–276.

Gunashekar, D.D., Bielak, L., Hägele, L., Oerther, B., Benndorf, M., Grosu, A.L., Brox, T., Zamboglou, C., & Bock, M., Explainable AI for CNN-based prostate tumor segmentation in multi-parametric MRI correlated to whole mount histopathology. *Radiation Oncology, 17(1)*, 65.

Guo, Y., Yu, T., Wu, J., Wang, Y., Wan, S., Zheng, J., Fang, L., & Dai, Q. (2022). Artificial intelligence for metaverse: a framework. *CAAI Artificial Intelligence Research, 1(1)*, 54–67.

Hashemzehi, R., Mahdavi, S. J. S., Kheirabadi, M., & Kamel, S. R. (2020). Detection of brain tumors from MRI images base on deep learning using hybrid model CNN and NADE. *Biocybernetics and Biomedical Engineering, 40(3)*, 1225–1232.

Huang, B., Zhao, B., & Song, Y. (2018). Urban land-use mapping using a deep convolutional neural network with high spatial resolution multispectral remote sensing imagery. *Remote Sensing of Environment, 214*, 73–86.

Jain, V., Nankar, O., Jerrish, D. J., Gite, S., Patil, S., & Kotecha, K. (2021). A novel AI-based system for detection and severity prediction of dementia using MRI. *IEEE Access, 9*, 154324–154346.

Janowicz, K., Gao, S., McKenzie, G., Hu, Y., & Bhaduri, B. (2020). GeoAI: spatially explicit artificial intelligence techniques for geographic knowledge discovery and beyond. *International Journal of Geographical Information Science, 34(4)*, 625–636.

Kuipers, B., & Byun, Y. T. (1991). A robot exploration and mapping strategy based on a semantic hierarchy of spatial representations. *Robotics and Autonomous Systems, 8(1–2)*, 47–63.

Law, S., Seresinhe, C. I., Shen, Y., & Gutierrez-Roig, M. (2020). Street-frontage-net: urban image classification using deep convolutional neural networks. *International Journal of Geographical Information Science, 34(4)*, 681–707.

Li, W., & Hsu, C. Y. (2020). Automated terrain feature identification from remote sensing imagery: a deep learning approach. *International Journal of Geographical Information Science, 34(4)*, 637–660.

Lo Gullo, R., Brunekreef, J., Marcus, E., Han, L.K., Eskreis-Winkler, S., Thakur, S.B., Mann, R., Groot Lipman, K., Teuwen, J. and Pinker, K. (2024). AI Applications to Breast MRI: Today and Tomorrow. *Journal of Magnetic Resonance Imaging*, 05 April.

Moldovan, D., Copil, G., & Dustdar, S. (2018). Elastic systems: Towards cyber-physical ecosystems of people, processes, and things. *Computer Standards & Interfaces., 57*, 76–82.

Openshaw, S. (1992). Some Suggestions concerning the development of artificial intelligence tools for spatial modelling and analysis in GIS. *Annals of Regional Science, 26(1)*, 35–51.

Openshaw, S., & Openshaw, C. (1997). *Artificial intelligence in geography*. John Wiley & Sons.

Papadimitriou, F. (2022a). Spatial entropy of landscapes simulated with artificial life and swarm intelligence. *Spatial Entropy and Landscape Analysis* (pp. 57–73). Springer VS.

Papadimitriou, F. (2022b). Emergence, sustainability and cyber-physical landscapes. *Spatial Entropy and Landscape Analysis* (pp. 123–139). Springer VS.

Papadimitriou, F. (2022c). Spatial negentropy and social self-organization in simulated landscapes. *Spatial Entropy and Landscape Analysis* (pp. 75–86). Springer VS.

Scott, G. J., England, M. R., Starms, W. A., Marcum, R. A., & Davis, C. H. (2017). Training deep convolutional neural networks for land–cover classification of high-resolution imagery. *IEEE Geoscience and Remote Sensing Letters, 14*(4), 549–553.

Smith, T. R. (1984). Artificial intelligence and its applicability to geographical problem solving. *Professional Geographer, 36*(2), 147–158.

Touya, G., Zhang, X., & Lokhat, I. (2019). Is deep learning the new agent for map generalization? *International Journal of Cartography, 5.2*(3), 142–157.

Van Eck, N. J., & Waltman, L. (2010). Software survey: VOS viewer, a computer program for bibliometric mapping. *Scientometrics, 84*(2), 523–538.

Wong, L. M., King, A. D., Ai, Q. Y. H., Lam, W. J., Poon, D. M., Ma, B. B., Chan, K. A., & Mo, F. K. (2021). Convolutional neural network for discriminating nasopharyngeal carcinoma and benign hyperplasia on MRI. *European Radiology, 31*, 3856–3863.

Zender, H., Mozos, O. M., Jensfelt, P., Kruijff, G. J., & Burgard, W. (2008). Conceptual spatial representations for indoor mobile robots. *Robotics and Autonomous Systems, 56*(6), 493–502.

Zhang, C., Sargent, I., Pan, X., et al. (2019a). Joint Deep Learning for Land Cover and Land Use Classification. *Remote Sensing of Environment, 221*, 173–187.

Zhang, S., Zhao, B., Tian, Y., & Chen, S. (2020). Stand with #StandingRock: Envisioning an epistemological shift in understanding geospatial big data in the "post-truth" era. *Annals of the American Association of Geographers*, 1–21.

Zhang, X., Karaman, S., & Chang, S. F. (2019). Detecting and simulating artifacts in GAN fake images. In *2019 IEEE international Workshop on Information Forensics and Security (WIFS)* (pp. 1–6).

Zhao, B., & Sui, D. Z. (2017). True lies in geospatial big data: Detecting location spoofing in social media. *Annals of GIS, 23*(1), 1–14.

Zhao, B., Zhang, S., Xu, C., Sun, Y., & Deng, C. (2021). Deep fake geography? When geospatial data encounter Artificial Intelligence. *Cartography and Geographic Information Science, 48*(4), 338–352.

Zhu, X. X., Tuia, D., Mou, L., et al. (2017). Deep learning in remote sensing: A comprehensive review and list of resources. *IEEE Geoscience and Remote Sensing Magazine, 5*(4), 8–3.

Chapter 2
Spatial AI in Symbolic and Evolutionary AI

Abstract This chapter examines how modelling and processing data describing entities of geographical interest such as land use and landscapes can lead to meaningful results for both humans and machines by using methods of Symbolic AI. Next, from building a simple expert system in Symbolic AI for assessing the outcomes of landscape change to tackling the Traveling Salesman Problem with the Ant Colony Optimization algorithm, it is shown how concepts and methods of Evolutionary AI (Artificial Swarm Intelligence and Agent-Based Models in particular) can serve in the creation of spatially-enabled AI that can help in spatial problem solving.

Keywords Spatial AI · Symbolic AI · Artificial swarm intelligence · Evolutionary AI · Agent-based models

2.1 Symbolic AI

Symbolic AI is endowed with the unique advantage of making understandable the communication between human thought and computers. The very first applications of AI in geography that prevailed in the 1990s and well into the first decade of the twenty-first century mainly focused on devising expert systems that were used for providing expert advice to humans. Those had mostly been based on Symbolic AI with expert systems exploiting languages that are understandable by both humans and machines. One such language is Prolog.

To illustrate the usefulness of Symbolic AI in representing spatial processes in a version of Prolog, an example of a simple AI expert system is presented here, that is suitable for describing certain geo-environmental impacts of landscape transformations. In order to create such a system, it is essential to draw from expertise that is already available about the spatial data and processes it relates to. In this case here, the spatial data relate to the dynamics of Southern-European landscapes (Papadimitriou and Mairota 1996; Papadimitriou 2023a, 2023b, 2023c) and are represented by three sets of variables: causes of landscape changes (c), landscape transformations (t) and effects of these transformations (e). In this small expert system, five causes of

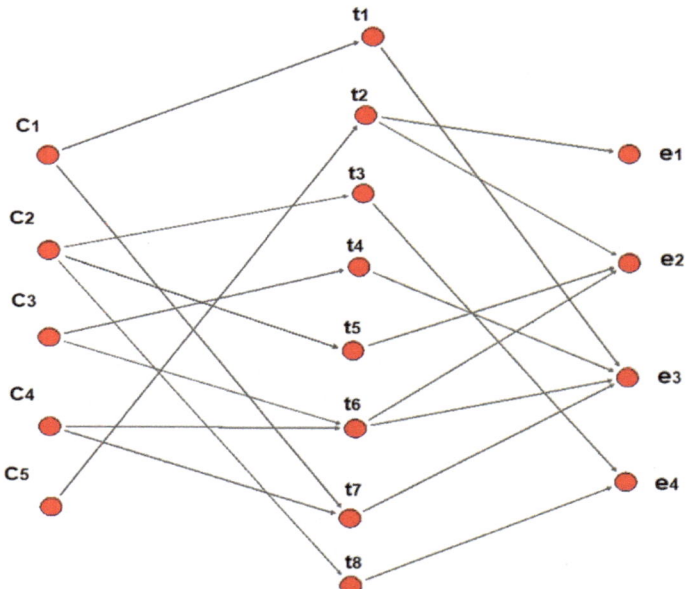

Fig. 2.1 The network of causes (c), transformations (t) and effects of transformations (e) modelled by an expert system written in Prolog

landscape change are considered only (c1 to c5) that trigger eight types of landscape transformations (t1 to t8), which, in turn, are responsible for four types of effects (e1 to e4) (Fig. 2.1).

The system was designed to be able to calculate the time (relative or real) that is required for a given effect to appear on the face of a landscape after certain "causes" have acted on it. Since each cause requires some time (CTW) to set off a landscape transformation in motion and since each effect may occur as soon as CTW is over, the realisation time (W) of the effect is calculated from CTW and the characteristic time of each effect (ETW). As the same effect may result from different causes and may be linked to more than one landscape transformations with different time values, this expert system is programmed to display the effects with their minimum values. Here is the central part of the system's database and the inference engine:

```
process_landscape_changes:-
landscape_change(C),
process_a_landscape_change(C),
fail.
process_landscape_changes.
```

```
process_a_landscape_change(C):-
landscape_change_transformation(C,T,CTW),
determine_ effects(C,T,CTW),
fail.
process_a_landscape_change(C).


determine_ effects(C,T,CTW):-
effect_transformation(E,T,ETW),
W is CTW + ETW,
check_existing_ effects(E,W),
fail.
determine_ effects(C,T,CTW).


check_existing_ effects(E,W):-
impact(E,OLD_W),
OLD_W >= W,!.
check_existing_ effects(E,W):-
not(effects(E,OLD_W)),
add_clause(effects(E,W)),!


display_result:-
write("The effects with their realization times are:"),
nl,write("Quickest:6, Slowest:1"),nl,
effect(E,W),
effect_name(E,N),
write(N),write(W),
nl,
fail.
display_result.
landscape_change_transformation database
landscape_change_transformation(c1,t7,3).
landscape_change_transformation(c1,t1,3).
landscape_change_transformation(c2,t3,1).
landscape_change_transformation(c2,t8,1).
landscape_change_transformation(c3,t6,1).
landscape_change_transformation(c3,t4,3).
landscape_change_transformation(c4,t6,2).
landscape_change_transformation(c4,t7,3).
landscape_change_transformation(c5,t2,3).
landscape_change_transformation(t1,expansion of residential land
use on agricultural areas).
landscape_change_transformation(t2,conversion   of   agriculture
into shrubland).
landscape_change_transformation(t3,conversion   of   agricultural
lands into bare ground).
landscape_change_transformation(t4,conversion   of   agriculture
into other land use).
```

```
landscape_change_transformation(t5,afforestation of agricultural
lands).
landscape_change_transformation(t6,agricultural   expansion   on
shrablands).
landscape_change_transformation(t7,   conversion   of   forest   to
shrubland).
landscape_change_transformation(t8,  reclamation  of  bare  ground
for agriculture).
effect-transformation database
effect_transformation(e1,t2,2).
effect_transformation(e2,t6,2).
effect_transformation(e2,t5,3).
effect_transformation(e2,t2,1).
effect_transformation(e3,t7,3).
effect_transformation(e3,t1,3).
effect_transformation(e3,t6,2).
effect_transformation(e3,t4,2).
effect_transformation(e4,t3,2).
effect_transformation(e4,t8,2).


landscape_change_name(c1,urbanization).
landscape_change_name(c2,afforestation).
Landscape_change_name(c3,land_reclamation_for_farming).
landscape_change_name(c4,deforestation).
landscape_change_name(c5,land_abandonement).


effect_name(e1,land_depreciation).
effect_name(e2,aesthetic_degradation).
effect_name(e3,increase_in_land_price).
effect_name(e4,aesthetic_improvement).
```

Given these, an example run of this program for the query of what are the effects of the landscape changes "deforestation" and "land reclamation for agriculture":

```
landscape_changes
deforestation
land_reclamation_for_agriculture
```

yields the following response by the system:

```
The effects with their realization times are:


Quickest:6
Slowest:1


increase in land price 6
aesthetic degradation 4
```

Provided that a suitable Prolog compiler has been installed on the computer, this code needs no more information than the above to deliver this response which resembles very much the way a human would answer based on the data and rules that enable one to infer from that data. Following this program, there may have been various outputs of "aesthetic degradation", but since the system displays a certain value for it (which is 4 in this case), it means all the other such effects had longer realization times (higher values): it is thus programmed to display the first appearance of an effect on the landscape rather than the last one. Effects that may conflict one another may well come out from within the same set of landscape responses and this possibility corresponds to what happens in geographical reality (different landscape transformations may simultaneously produce different effects in different parts of the geographical region that is studied). Thus, a program written in an AI language with only a few lines of code enables a computer to "understand" queries about spatial changes that are expressed in human language and provide numerical estimates that are written in a way that is understandable by humans.

2.2 Evolutionary AI

Besides Symbolic AI, the growth of Evolutionary AI has revealed applications that are explicitly spatial. Artificial Swarm Intelligence is one of them and it is an important component of AI (Jangra et al., 2013; Kennedy, 2006; Iba, 2019; Rosenberg and Wilcox, 2020; Arvidsson, 2020) with its theory and methods (Rosenberg, 2016a) articulated around the theory of Swarm Intelligence (Bansal et al., 2019; Bonabeau et al., 1999; Chakraborty & Kar, 2017; Nayyar & Nguyen, 2018) and with numerous applications in diverse domains such as swarm robotics (Cai et al., 2023; Sharkey, 2006), UAV swarms (Johnson, 2020; Puente-Castro et al., 2022; Spanaki et al., 2022), entertainment (Roy et al. 2014) etc. Artificial Swarm Intelligence can only be perceived in a spatial context as swarms display their remarkable self-organization behaviors in space (Theraulaz et al., 2002), both in 2D (i.e. ants) and in 3D (i.e. flocks of birds and fish).

Swarm Intelligence exploits the phenomenon of stigmergy that is observed in the animal kingdom, by which individuals appear as if they were alone, while they act in a coordinated manner within the framework of some collective activity (Theraulaz & Bonabeau, 1999). Ant colonies and bee hives are prominent examples of stigmergy in nature, whence the Ant Colony Optimization-ACO (Fig. 2.2) and the "bee hive" algorithms (Fig. 2.3) that mimic the behavior of these insects and have contributed to the rise of swarm intelligence (Roy et al., 2014; Papadimitriou, 2022a, 2022b, 2022c). Interestingly, the exploitation of spatial features of a variety of combinations of ABMs, swarm intelligence and CNNs has found multiple applications in image steganography (Shokranipour and Hasanzadeh 2016; Chaumont 2019; Duan et al.

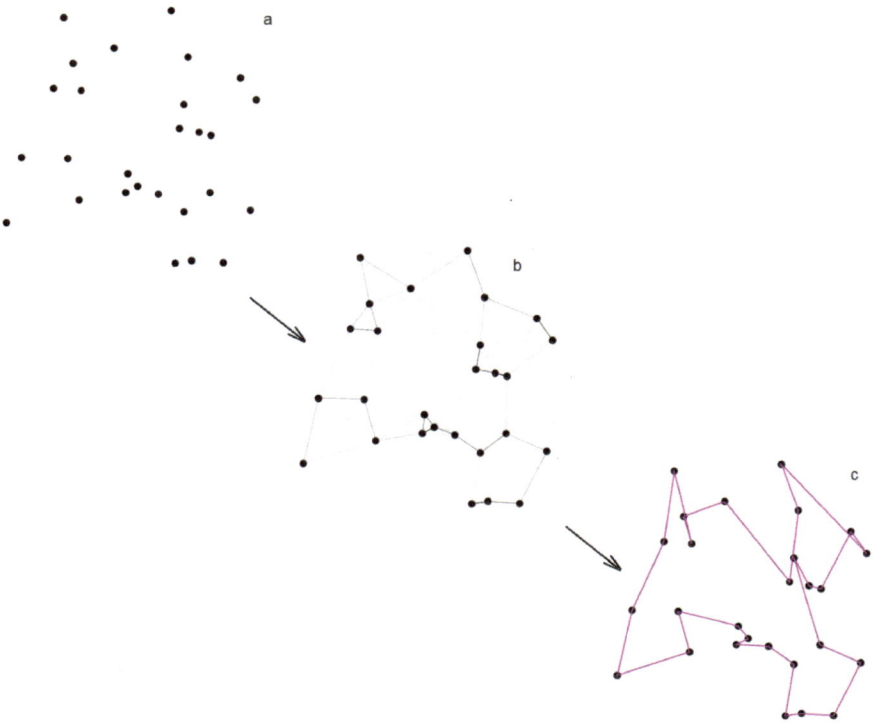

Fig. 2.2 Stages of tackling the traveling salesman problem by means of the ACO algorithm: from an initial set of points (**a**) possible links are tried out (**b**) until a solution is reached (**c**)

2020; Grurunath et al. 2021; Zebari et al. 2020; Vazquez et al. 2022; Wani and Sultan 2023).

A characteristic example of their importance in spatial analysis is that ACO algorithms can yield near-optimal solutions to the "Traveling Salesman Problem". The key idea in ACO is the "pheromone matrix" that models the amount of pheromone between two points in space. The entries of this matrix are updated once all points in the pre-defined spatial domain have been visited by the ant population, after all the iterations have been performed.

Stigmergy is frequently observed in swarm dynamics, governing the spatial behavior of virtual swarms of birds, such as "boids" (Fig. 2.4). Three types of interactions define the motion of boids: attraction (movement towards the centre of mass of their neighbours), repulsion (avoidance of neighbouring birds) and orientation (movement towards the same direction as their neighbors). Stigmergy also underlies the intersection of Spatial AI with robotics, whence the "swarm robots" (Blum & Groß, 2015; Fan et al., 2022).

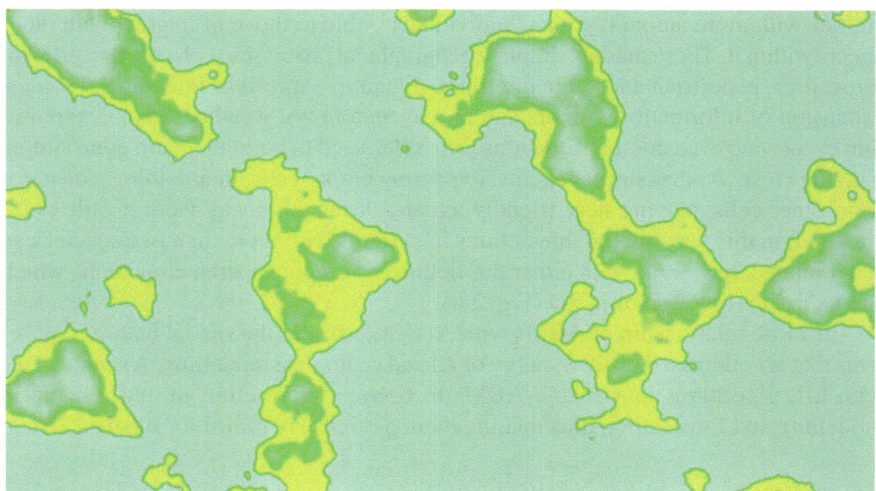

Fig. 2.3 Spatial search created by using the bee colony algorithm: more frequently visited areas appear in yellow

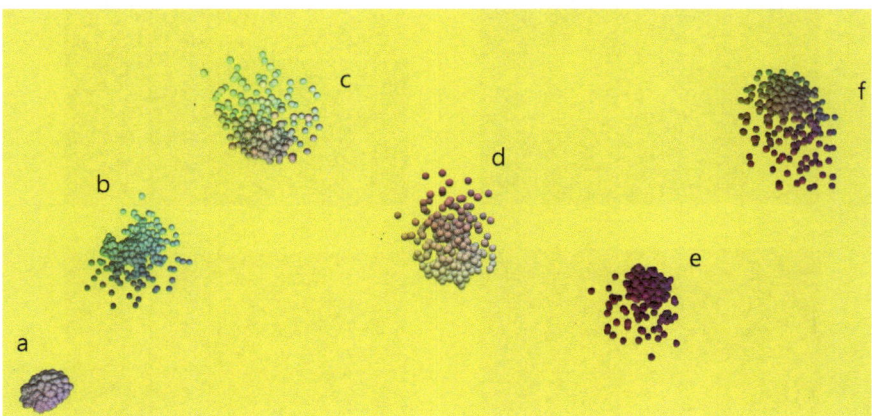

Fig. 2.4 Snapshots of an artificial swarm moving in digital space, changing directions of movement and exhibiting stigmergy and spatial self-organization (from a to f)

Besides swarm intelligence, Agent-Based Models (ABMs) constitute another class of spatial models that have been developed drawing from expertise accumulated from within a diverse array of disciplines and theories, such as cellular automata, percolation theory, lattice-gas models, symbolic dynamics, evolutionary dynamics etc. Modelling and simulation with ABMs is effectuated within well-defined spatial domains (most often square lattices). Each cell of the lattice is endowed with some prescribed range of possible behaviors and the "agents" are dynamic types of cellular entities with more autonomy than "individuals", able to move or interact with other agents within it. These models simulate geographical processes such as urban growth, forest fires, pedestrian and vehicular traffic dynamics, spread of infectious diseases, expansion of information, rumors and propaganda flows, weather dynamics, pollution events and even combat situations. The rules used to build them are quite simple (staying close to other similar agents, bypassing empty cells, maintaining a distance from other cells, staying near friendly agents, distancing away from hostile ones, etc.) and, in this way, spatial interactions in systems with more than two species can be simulated, beginning with either pre-defined or random spatial allocations which can be in 2D (Fig. 2.5) or in 3D (Fig. 2.6).

The close relationship of ABMs with AI demonstrates the spatial nature of problems that are addressed by the synergy of AI and ecological modelling. AI contributes with ML algorithms that enable ABMs to become more efficient and reliable in modelling and forecasting, thus making them particularly useful for medical, social

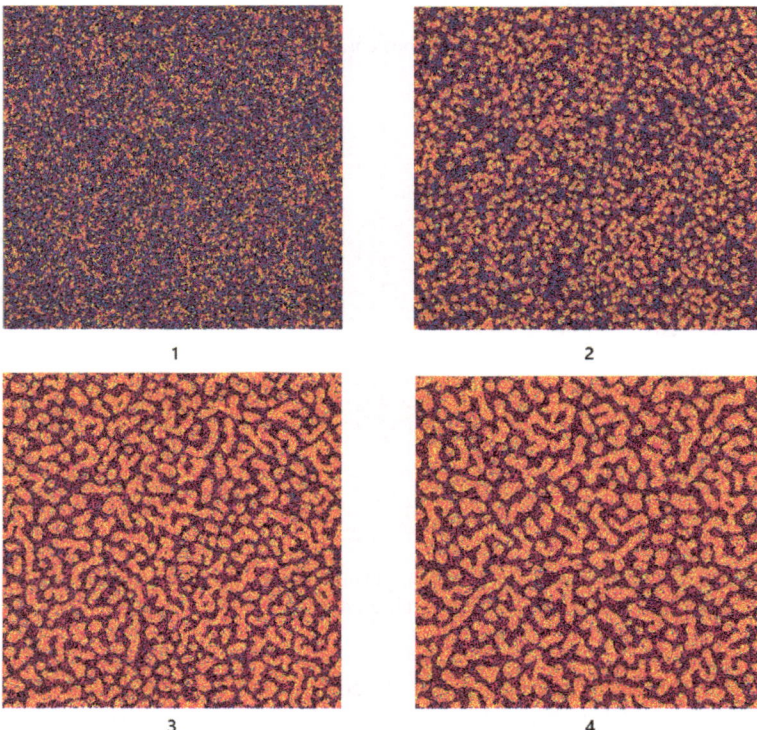

Fig. 2.5 Four stages of development of a 2D ABM model that is capable of simulating spatial aggregation in the course of time. The model depicts the interaction of three species in a 2D space that also contains empty cells (black). Beginning with a random allocation of species (1) and after some iterations (2–3), two species dominate the spatial domain (3), and, eventually, large aggregates of cells develop (4)

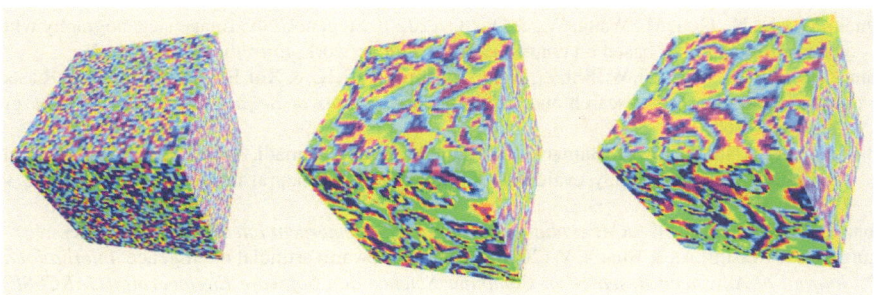

Fig. 2.6 An ABM model that is capable of producing aggregates in 3D space from a random spatial allocation (left) to the formation of large aggregates (right)

and ecological applications. As a result, theoretical problems of geography can be explored by using ABMs, i.e. explaining the rise, expansion and dissolution of spatial hierarchical structures, the development of non-local interactions, as well as the creation and transformation of spatial networks and the changes in flows of energy, matter and information in the geographical space. Among the most interesting future challenges at the intersection of AI with ABMs will be to model and simulate agent maneuvers and deception tactics with agents hiding from other agents or pursuing collective survival at the expense of individual goals (altruistic behaviors). In such cases, ABMs are expected to function in an intelligent way and thus further contribute to the predictive capacity of Spatial AI.

References

Arvidsson, M. (2020). The swarm that we already are: Artificially intelligent (AI) swarming 'insect drones', targeting and international humanitarian law in a posthuman ecology. *Journal of Human Rights and the Environment, 11*(1), 114–137.

Aviv, A. J., Budzitowski, D., & Kuber, R. (2015). Is bigger better? Comparing user-generated passwords on 3x3 vs. 4x4 grid sizes for Android's pattern unlock. In *Proceedings of the 31st annual computer security applications conference*, pp. 301–310.

Bansal, J. C., Singh, P. K., & Pal, N. R. (Eds.). (2019). *Evolutionary and swarm intelligence algorithms* (Vol. 779). Springer.

Blum, C., & Groß, R. (2015). Swarm intelligence in optimization and robotics. *Springer Handbook of Computational Intelligence* (pp. 1291–1309).

Bonabeau, E., Dorigo, M., & Theraulaz, G. (1999). *Swarm intelligence: From natural to artificial systems*. Oxford University Press.

Cai, W., Liu, Z., Zhang, M., & Wang, C. (2023). Cooperative artificial intelligence for underwater robotic swarm. *Robotics and Autonomous Systems, 164*, 104410.

Chakraborty, A., & Kar, A. K. (2017). Swarm intelligence: A review of algorithms. *Nature-inspired computing and optimization: Theory and applications*, 475–494.

Chaumont, M. (2019). Deep learning in steganography and steganalysis from 2015 to 2018. ArXiv: 1904.01444.

Duan, X., Liu, N., Gou, M., Wang, W., & Qin, C. (2020). SteganoCNN: Image steganography with generalization ability based on convolutional neural network. *Entropy, 22*(10), 1140.

Fan, Z., Sun, F., M. A., P., Li, W., Shi, Z., Wang, Z., Zhu, G., Li, K., & Xin, B.(2022). Stigmergy-based swarm robots for target search and trapping. *Transactions of Beijing institute of Technology, 42*(2), 158–167.

Gurunath, R., Alahmadi, A. H., Samanta, D., Khan, M. Z., & Alahmadi, A. (2021). A novel approach for linguistic steganography evaluation based on artificial neural networks. *IEEE Access, 9*, 120869–120879.

Iba, H. (2019). *AI and SWARM: evolutionary approach to emergent intelligence.* CRC Press.

Jangra, A., Awasthi, A., & Bhatia, V. (2013). A study on swarm artificial intelligence. *International Journal of Advanced Research in Computer Science and Software Engineering (IJARCSSE), 9*(8).

Johnson, J. (2020). Artificial intelligence, drone swarming and escalation risks in future warfare. *The RUSI Journal, 165*(2), 26–36.

Kennedy, J. (2006). Swarm intelligence. *Handbook of nature-inspired and innovative computing: Integrating classical models with emerging technologies* (pp. 187–219). Springer, US.

Nayyar, A., & Nguyen, N. G. (2018). Introduction to swarm intelligence. *Advances in swarm intelligence for optimizing problems in computer science* (pp. 53–78). Chapman and Hall.

Papadimitriou, F. (2012). Artificial Intelligence in modelling the complexity of Mediterranean landscape transformations. *Computers and Electronics in Agriculture, 81*, 87–96.

Papadimitriou, F. (2022a). *Spatial entropy and landscape analysis.* Springer VS.

Papadimitriou, F. (2022b). Spatial entropy of landscapes simulated with artificial life and swarm intelligence. *Spatial entropy and landscape analysis* (pp. 57–73). Springer VS.

Papadimitriou, F. (2022c). Spatial negentropy and social self-organization in simulated landscapes. *Spatial entropy and landscape analysis* (pp. 75–86). Springer VS.

Papadimitriou, F. (2022d). Emergence, sustainability and cyber-physical landscapes. *Spatial entropy and landscape analysis* (pp. 123–139). Springer VS.

Papadimitriou, F. (2023a). *Modelling landscape dynamics.* Springer VS.

Papadimitriou, F. (2023b). Dynamical systems modelling of landscape transformations. *Modelling Landscape Dynamics: Determinism, Stochasticity and Complexity* (pp. 1–15). Springer VS.

Papadimitriou, F. (2023c). Markov models of landscape dynamics. *Modelling Landscape Dynamics: Determinism, Stochasticity and Complexity* (pp. 45–57). Springer VS.

Papadimitriou, F., & Mairota, P. (1996). Spatial scale-dependent policy planning for land management in southern Europe. *Environmental Monitoring and Assessment, 39*, 47–57.

Puente-Castro, A., Rivero, D., Pazos, A., & Fernandez-Blanco, E. (2022). A review of artificial intelligence applied to path planning in UAV swarms. *Neural Computing and Applications, 34*(1), 153–170.

Rosenberg, L., & Willcox, G. (2020). Artificial swarm intelligence. In *Intelligent Systems and Applications: Proceedings of the 2019 Intelligent Systems Conference (IntelliSys)*(Vol. 1; pp. 1054–1070). Springer.

Roy, S., Biswas, S., & Chaudhuri, S. S. (2014). Nature-inspired swarm intelligence and its applications. *International Journal of Modern Education and Computer Science, 6*(12), 55.

Sharkey, A. J. (2006). Robots, insects and swarm intelligence. *Artificial Intelligence Review, 26*, 255–268.

Shokranipour, S., & Hasanzadeh, M. (2016). High capacity image steganography method by using particle swarm optimization. *The Modares Journal of Electrical Engineering, 15*(4), 1–7.

Spanaki, K., Karafili, E., Sivarajah, U., Despoudi, S., & Irani, Z. (2022). Artificial intelligence and food security: Swarm intelligence of AgriTech drones for smart AgriFood operations. *Production Planning and Control, 33*(16), 1498–1516.

Theraulaz, G., & Bonabeau, E. (1999). A brief history of stigmergy. *Artificial Life, 5*(2), 97–116.

Theraulaz, G., Bonabeau, E., Nicolis, S. C., Solé, R. V., Fourcassié, V., Blanco, S., Fournier, R., Joly, J. L., Fernández, P., Grimal, A., & Dalle, P. (2002). Spatial patterns in ant colonies. *Proceedings of the National Academy of Sciences, 99*(15), 9645–9649.

Vazquez, E., Torres, S., Sanchez, G., Avalos, J. G., Abarca, M., Frias, T., Juarez, E., Trejo, C., & Hernandez, D. (2022).Confidentiality in medical images through a genetic-based steganography algorithm in artificial intelligence. *Frontiers in Robotics and AI, 9*, 1031299.

Wani, M. A., & Sultan, B. (2023). Deep learning based image steganography: a review. *Wiley Interdisciplinary Reviews: Data Mining and Knowledge Discovery, 13*(3), e1481.

Zebari, D. A., Zeebaree, D. Q., Saeed, J. N., Zebari, N. A., & Adel, A. Z. (2020). Image steganography based on swarm intelligence algorithms: A survey. *People, 7*(8), 9.

Chapter 3
Spatial AI, Generative AI and ChatGPT

Abstract This chapter traces the developments in Spatial AI from Neuro-Symbolic to Generative AI, demonstrating how AI can be created and/or derived from spatial information and spatial data. Beginning with Google's Bert and ending with Generative AI and ChatGPT, it is shown how spatial or/and geographical information can be retrieved from within Large Language Models and how it can be represented, analysed and processed using convolutional neural networks and generative adversarial networks.

Keywords Spatial AI · Convolutional neural networks · GANs · Generative AI · ChatGPT · Neuro-symbolic AI · Bert · Google trends

3.1 Generative AI

Neuro-symbolic AI integrates symbolic AI with neural (connectionist) approaches (Hitzler et al., 2022) and there are several alternative ways by which symbolic AI can be endowed with neural networks in order to create different levels of neuro-symbolic AI (Sarker et al., 2021). While in some cases such hybrid networks belong to the class of Explainable Neural Networks (XNNs), in other cases they are not explainable but they can still be very useful. A prominent example of neuro-symbolic AI is Google's BERT, which, along with ChatGPT-3, is a Large Language Model (LLM) that uses such a hybrid technology. With the exception of ChatGPT, perhaps no other widely known example of how natural language processing (NLP) algorithms can be made easily accessible and useful is more characteristic than "Google Trends". Google's AI algorithm "BERT" (an acronym for Bidirectional Encoder Representations from Transformers) is able to "understand" the context of a word by taking into account both its preceding and subsequent words (Devlin et al., 2018; Singh, 2021). It is called "bidirectional" because it encodes sentences in both directions, "encoder representation" because it is able to translate word expressions in a way that it is understandable by its algorithms and "transformer" as it encodes the words of each sentence so that they become meaningful by using NLP algorithms. The latter are machine learning

F. Papadimitriou, *Spatial Artificial Intelligence*,
SpringerBriefs in Computational Intelligence,
https://doi.org/10.1007/978-3-031-82136-3_3

techniques that are also used in other widely known applications such as "Grammarly", "Alexa", "Cortana" etc. and are able to perform various analyses of natural language, also with respect to language syntax (see Ullah et al., 2023). Endowed with these properties and functions, BERT is an automatic supervised text classification system of AI that does not require human supervision (González-Carvajal & Garrido-Merchán, 2020), able to search for possible relationships among words, analyzing the resulting meanings and eventually coming up with expressions that resemble the human way of understanding sentences (see Wiedemann et al., 2019; Hao et al., 2019; Rogers et al., 2020). BERT can be used to explore spatial information also (Shin et al., 2020; Sigirci & Bilgin, 2022; Yang et al., 2022), i.e. map representations of Google searches as it is in-built in Google Trends. Indispensable in analyses of public interest in any word or term (from any place of the world and anytime from 2004 onwards), Google Trends provides normalized counts of searches (on a scale from 0 to 100) for any word or combination of words, by generating a time series of Search Volume Indices (SVIs) for any search term. Depending on the desired resolution in time, the series can be daily, weekly or monthly data. For multiple terms, it generates an SVI for each term, which is subsequently normalized with respect to the term with the highest SVI. Google Trends can be easily used to delve into information that subsequently can be rendered spatial by providing world maps of word searches. For instance, using Google Trends enables us to derive spatial visualizations of the total searches for two terms that are of wide interest: "sustainability" and "resilience", both at country level and worldwide, for the period 2004–2024. For either one (or both), it can provide maps according to the numbers of Google searches of that term(s) at the country level. In this way, the prevailing geographical differences in the spatial distribution of Google Trends searches of these terms can be visualized as maps (Fig. 3.1) informing us i.e. that people from North America, east African countries and Oceania have been more concerned about sustainability rather than people from the European countries, while searches for resilience prevailed over those for sustainability in two countries (France and Nepal).

Fig. 3.1 Maps of searches for the terms sustainability and resilience worldwide, from 2004 until 2024. The prevailing color per country indicates more searches for one term at the expense of the other. Snapshots taken on 14 October 2024. *Data source* Google Trends (https://www.google.com/trends)

Perhaps the most striking manifestation of the current explosive growth in Spatial AI is due to convolutional neural networks (CNNs) and generative adversarial networks (GANs). The form of a typical neural network (Fig. 3.2) can be enriched with more layers of nodes and more links among nodes, if the nodes are images and the operation that connects the layers is convolution of images (Fig. 3.3), followed by max-pooling (Fig. 3.4), eventually leading to CNNs (Fig. 3.5).

GANs expand the scope of CNNs by using two"adversaries": a "generator" and a "discriminator"- each one of these being a set of images. The former is trained to generate maps from an initial dataset, while the latter receives the result and evaluates it. Consequently, the output that the discriminator generates is fed into the generator and, in this way, the generator produces increasingly more realistic images. Although CNNs can be programmed by using PyTorch, Python, Keras etc., a wide range of different models can be used for GenAI: VariationalAutoEncoders (VAEs), GANs, Transformers, Language Models, as well as hybrid models, while the procedures by which GANs can be trained can be refined by means of "Progressive GANs" that increase the accuracy of automatic classification (Collier et al., 2018).

There are two main categories of AI for image generation: text-to-image and image-to-image. Models for transforming a text to an image exploit GANs (Zhang

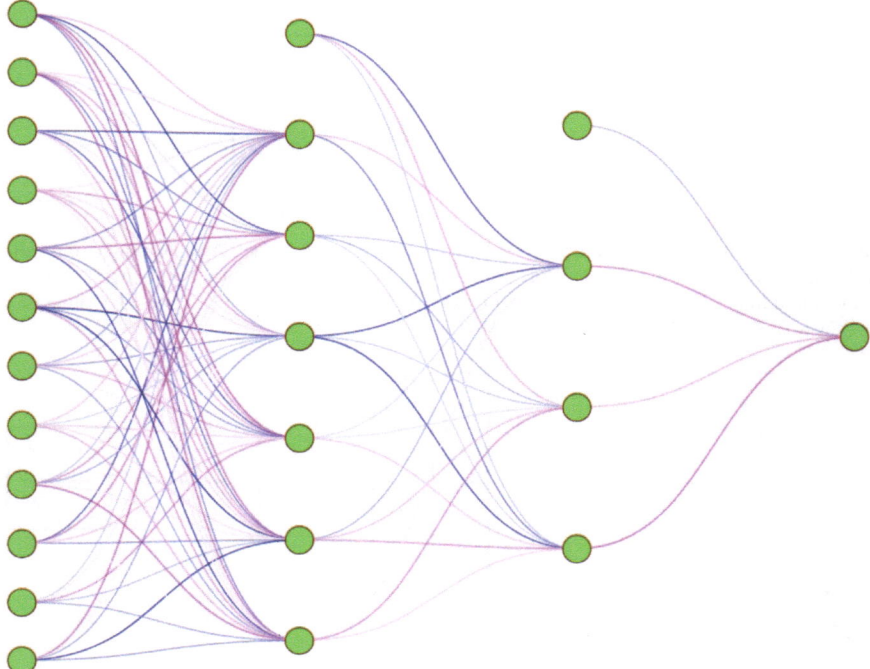

Fig. 3.2 A neural network with four layers: an input layer with 12 nodes, two hidden layers with 7 and 4 nodes respectively, and an output layer with one node

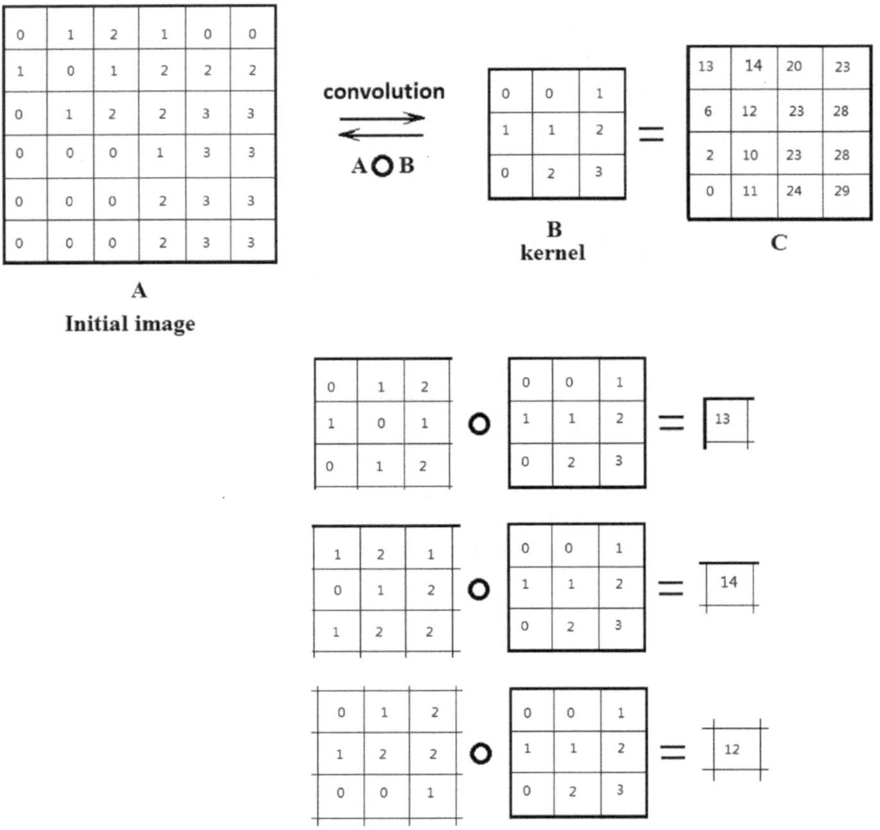

0	1	2	1	0	0
1	0	1	2	2	2
0	1	2	2	3	3
0	0	0	1	3	3
0	0	0	2	3	3
0	0	0	2	3	3

A
Initial image

convolution
$\rightleftarrows$
A O B

B kernel

0	0	1
1	1	2
0	2	3

=

13	14	20	23
6	12	23	28
2	10	23	28
0	11	24	29

C

0	1	2
1	0	1
0	1	2

O

0	0	1
1	1	2
0	2	3

= 13

1	2	1
0	1	2
1	2	2

O

0	0	1
1	1	2
0	2	3

= 14

0	1	2
1	2	2
0	0	1

O

0	0	1
1	1	2
0	2	3

= 12

Fig. 3.3 The convolution of an initial 6 × 6 image with a 3 × 3 kernel. As an example, the value of the upper left cell of the image C has been derived from the convolution of the first 3 × 3 cells of the initial image with those of the kernel image: $(0 \times 0) + (1 \times 0) + (2 \times 1) + (1 \times 1) + (0 \times 1) + (1 \times 2) + (0 \times 0) + (1 \times 2) + (2 \times 3) = 13$

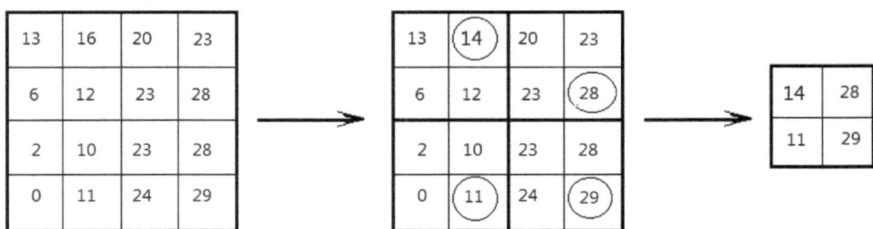

Fig. 3.4 The max-pooling procedure

Fig. 3.5 A CNN created from a stack of 26 images of size 117×113 after convoluting them with other images and max-pooling

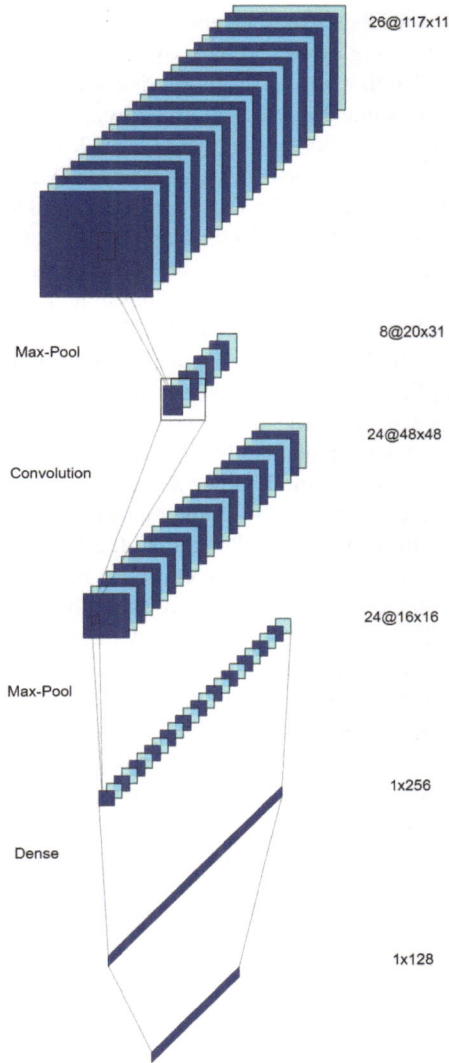

26@117x113

Max-Pool

8@20x31

Convolution

24@48x48

Max-Pool

24@16x16

1x256

Dense

1x128

et al., 2017) and diffusion models (Nichol et al., 2021). Creating images from other images (image-to-image translation) is also possible by means of both GANs (Chen & Hays, 2018) and diffusion models (Peng et al., 2023), although by adopting different methods of data processing. The main disadvantages of GANs are their instability in training, the "mode collapse" and the large data sets required to train them. With the alternative technology of "Generative Diffusion Models" (GDMs), new images can be generated by starting from noise and ridding of it gradually. Among the disadvantages of such models are their requirement for larger datasets, the longer processing times and their overly complex resulting structures.

The first and most impressive applications of GANs in image analysis (Arhcana Balkrishna, 2024; Casteleiro-Pitrez, 2024; Zhu et al., 2024; Zhou & Lee, 2024) have been in medicine (Chen et al., 2022; Li et al., 2021) and geography, in which they opened up the possibility to create truly astounding fake geographic representations (Abady et al., 2022; Gil-Fournier & Parikka, 2021; Zhao et al., 2021) with stunning resemblance to real landscapes, maps or satellite images. Yet, the spatial context of these models needn't be confined to two dimensions only; converting images to 3D representations is also possible using software such as "Magic123", "Dream-Craft3D", "NeRF", "CLIP-NeRF" etc., while there are also 3D models of GenAI that are suitable for designing avatars (Zhang et al., 2023a, 2023b). Further, besides its applications in image processing and geography, GAN-based Spatial AI turned out to be useful in the automatic creation of architectural plans and 3D urban spatial data that can be generated by GANs (Hu et al., 2024) and the concept of "Image Generative Artificial Intelligence Systems" (IGAI) has been proposed (Guridi et al., 2024), whereas Large-scale Language Models (LLMs) can be adjusted so as to become useful for IGAI (Wideinger et al. 2022).

Linking LLMs with GenAI can be a particularly promising prospect, since input from GPT can serve to derive spatially-enbled output (Li & Ning, 2023; Zhang et al., 2023a). Image-to-image GANs transform images to other images (Taigman et al., 2016; Kim et al., 2017), while text-to-image GANs produce images on the basis of text prompts (Zhang et al., 2017). Alternatively, diffusion models insert diffusion into images until the original spatial data have been altered enough so as to conform with a predefined distribution and consequently, by learning this gradual degradation process, the diffusion model can reverse the procedure by removing the noise and then derive a realistic output that resembles the original image (of particular interest in this respect are hybrid models, i.e. GANs-and-Diffusion, CNNs-and-Transformers etc). Expanding on these, combinations of image-and-text as input to derive images as output are also possible with a variety of models that entail GANs or diffusion.

3.2 ChatGPT

The very popular ChatGPTs belong to the Generative Pre-trained Transformer (GPT) models of LLMs (Wu et al., 2023). ChatGPT-3.5 was launched in late 2022 and was followed by Bard (Gemini), Bing (Copilot), Claude-2 and GPT-4. ChatGPTs can generate codes in several programming languages (i.e. C++, Python, Java). It may come as a surprise that an LLM such as ChatGPT can create maps (with GeoPANDAs) or render shapefiles into other forms (e.g. GeoJSON, GDAL). The geographical information can be processed as JSON files, whereas the procedures are handled using Python libraries (e.g. Matplotlib) and PyQGIS scripts (Hochmair et al., 2024). Geographical queries can be expressed in ChatGPT using SPARQL, while the models draw data from Dbpedia, Mapbox9, Open Street Map7, World

Map10, Google Maps8, D3js11. Consequently, LLMs provide answers to geographical questions relating to toponyms, time series and proximity and are able to perform basic GIS operations (Tao & Shu 2023; Zhang et al., 2024).

Besides their role as auxiliary systems for geographical education (Chang & Kidman, 2023), ChatGPTs are also useful in exploring the interactions of humans with computer creativity in terms of spatial thinking, geoparsing and map design (Ghanim, 2024; Li & Ning, 2023). Visual ChatGPT for instance, can generate textual descriptions for satellite images and can perform some basic image processing operations such as edge detection and segmentation (Osco et al., 2023). Whichever particular model is adopted to advance such forms of Spatial AI, it is important to notice that ChatGPT-4 presents a significant advance with respect to ChatGPT-3.5, since the former can process both textual and visual inputs. So long as ChatGPTs are able to access large and qualitatively adequate datasets related to the prompt or question received from the user, they are expected to be able to answer with high accuracy and reliability. However, this is not always the case and, failing one (or both) of the above conditions may lead ChatGPTs to provide wrong answers, which, in turn, may compromise the reliability of the information they provide, possibly also raising ethical concerns along with the erroneous outputs.

As a further development in Spatial AI, LLMs, transformers as well as diffusion models are used to create generative video (Hong et al., 2022; Ho et al., 2022; Kondratyuk et al., 2023; Gupta et al., 2023). Actually, both OpenAI's "Sora" and Google's "Lumiere" transform text or images into videos. "Sora" compresses videos and turns them into spacetime patches in a space of lower dimension (Liu et al., 2024; Mogavi et al., 2024; Wang et al., 2024; Zhou et al., 2024). "Lumiere" uses the "Space–Time U-Net" architecture that is able to convert the original (prompt) image into a video animation (Bar-Tal et al., 2024). To achieve this, millions of videos along with their captions are processed in a model trained at 128×128 resolution, while the generated videos are 80 frames long and are reproduced at a rate of 16 frames per second.

References

Abady, L., Horváth, J., Tondi, B., Delp, E. J., & Barni, M. (2022). Manipulation and generation of synthetic satellite images using deep learning models. *Journal of Applied Remote Sensing, 16*(4), 046504.

Archana Balkrishna, Y. (2024). An analysis on the use of image design with generative AI technologies. *International Journal of Trend in Scientific Research and Development, 8*(1), 596–599.

Bar-Tal, O., Chefer, H., Tov, O., Herrmann, C., Paiss, R., Zada, S., Ephrat, A., Hur, J., Liu, G., Raj, A., & Li, Y. (2024). Lumiere: A space-time diffusion model for video generation. arXiv: 2401.12945.

Casteleiro-Pitrez, J. (2024). Generative artificial intelligence image tools among future designers: A usability, user experience, and emotional analysis. *Digital, 4*(2), 316–332.

Chang, C. H., & Kidman, G. (2023). The rise of generative artificial intelligence (AI) language models-challenges and opportunities for geographical and environmental education. *International Research in Geographical and Environmental Education, 32*(2), 85–89.

Chen, Y., Yang, X. H., Wei, Z., Heidari, A. A., Zheng, N., Li, Z., Chen, H., Hu, H., Zhou, Q., & Guan, Q. (2022). Generative adversarial networks in medical image augmentation: A review. *Computers in Biology and Medicine, 144*, 105382.

Chen, W., & Hays, J. (2018). Sketchygan: Towards diverse and realistic sketch to image synthesis. In *Proceedings of the IEEE Conference on Computer Vision and Pattern Recognition* (pp 9416–9425).

Collier, E., Duffy, K., Ganguly, S., Madanguit, G., Kalia, S., Shreekant, G., Nemani, R., Michaelis, A., Li, S., Ganguly, A. & Mukhopadhyay, S. (2018). Progressively growing generative adversarial networks for high resolution semantic segmentation of satellite images. In *2018 IEEE International Conference on Data Mining Workshops (ICDMW)* (pp. 763–769). IEEE.

Devlin, J., Chang, M. W., Lee, K., & Toutanova, K. (2018). Bert: Pre-training of deep bidirectional transformers for language understanding. arXiv: 1810.04805.

Ghanim, H. (2024). Artificial intelligence and the spatial documentation of languages. arXiv: 2404.01263.

Gil-Fournier, A., & Parikka, J. (2021). Ground truth to fake geographies: Machine vision and learning in visual practices. *AI & Society, 36*(4), 1253–1262.

González-Carvajal, S., & Garrido-Merchán, E. C. (2020). Comparing BERT against traditional machine learning text classification. arXiv: 2005.13012.

Gupta, A., Yu, L., Sohn, K., Gu, X., Hahn, M., Li, F.F., Essa, I., Jiang, L., & Lezama, J. (2023). Photorealistic video generation with diffusion models. arXiv: 2312.06662.

Guridi, J. A., Cheyre, C., Goula, M., Santo, D., Humphreys, L., Shankar, A., & Souras, A. (2024). Image generative AI to design public spaces: A reflection of how AI could improve co-design of public parks. *Digital Government: Research and Practice.*

Hao, Y., Dong, L., Wei, F., & Xu, K. (2019). Visualizing and understanding the effectiveness of BERT. arXiv: 1908.05620.

Hitzler, P., Eberhart, A., Ebrahimi, M., Sarker, M. K., & Zhou, L. (2022). Neuro-symbolic approaches in artificial intelligence. *National Science Review, 9*(6), nwac035.

Ho, J., Chan, W., Saharia, C., Whang, J., Gao, R., Gritsenko, A., Kingma, D. P., Poole, B., Norouzi, M., Fleet, D. J., & Salimans, T. (2022). Imagen video: High definition video generation with diffusion models. arXiv: 2210.02303.

Hochmair, H. H., Juhasz, L., & Kemp, T. (2024). Correctness comparison of ChatGPT-4, Bard, Claude-2, and Copilot for Spatial Tasks. arXiv: 2401.02404.

Hong, W., Ding, M., Zheng, W., Liu, X., & Tang, J. (2022). Cogvideo: Large-scale pretraining for text-to-video generation via transformers. arXiv: 2205.15868.

Hu, X., Lin, C., Chen, T., & Chen, W. (2024). Interactive design generation and optimization from generative adversarial networks in spatial computing. *Scientific Reports, 14*(1), 5154.

Kondratyuk, D., Yu, L., Gu, X., Lezama, J., Huang, J., Schindler, G., Hornung, R., Birodkar, V., Yan, J., Chiu, M. C. & Somandepalli, K. (2023). Videopoet: A large language model for zero-shot video generation. arXiv: 2312.14125.

Li, X., Jiang, Y., Rodriguez-Andina, J. J., Luo, H., Yin, S., & Kaynak, O. (2021). When medical images meet generative adversarial network: Recent development and research opportunities. *Discover Artificial Intelligence, 1*, 1–20.

Li, Z., & Ning, H. (2023). Autonomous GIS: The next-generation AI-powered GIS. arXiv: 2305.06453.

Liu, Y., Zhang, K., Li, Y., Yan, Z., Gao, C., Chen, R., Yuan, Z., Huang, Y., Sun, H., Gao, J., & He, L. (2024). Sora: A review on background, technology, limitations, and opportunities of large vision models. arXiv: 2402.17177.

Mogavi, R. H., Wang, D., Tu, J., Hadan, H., Sgandurra, S. A., Hui, P., & Nacke, L. E. (2024). Sora OpenAI's Prelude: Social Media Perspectives on Sora OpenAI and the Future of AI Video Generation. arXiv: 2403.14665.

Nichol, A., Dhariwal, P., Ramesh, A., Shyam, P., Mishkin, P., McGrew, B., Sutskever, I., & Chen, M. (2021). Glide: Towards photorealistic image generation and editing with text-guided diffusion models. arXiv: 2112.10741.

Osco, L. P., Lemos, E. L. D., Gonçalves, W. N., Ramos, A. P. M., & Marcato Junior, J. (2023). The potential of visual ChatGPT for remote sensing. *Remote Sensing, 15*(13), 3232.

Peng, Y., Zhao, C., Xie, H., Fukusato, T., & Miyata, K. (2023). Difffacesketch: High-fidelity face image synthesis with sketch-guided latent diffusion model. arXiv: 2302.06908.

Rogers, A., Kovaleva, O., & Rumshisky, A. (2020). A primer in BERTology: What we know about how BERT works. *Transactions of the Association for Computational Linguistics, 8*, 842–866.

Sarker, M. K., Zhou, L., Eberhart, A., & Hitzler, P. (2021). Neuro-symbolic artificial intelligence. *AI Communications, 34*(3), 197–209.

Shin, H. J., Park, J. Y., Yuk, D. B., & Lee, J. S. (2020). Bert-based spatial information extraction. In *Proceedings of the third international workshop on spatial language understanding* (pp. 10–17).

Sigirci, I. O., & Bilgin, G. (2022). Spectral-spatial classification of hyperspectral images using Bert-based methods with HyperSLIC segment embeddings. *IEEE Access, 10*, 79152–79164.

Singh, S. (2021). BERT algorithm used in google search. *Mathematical Statistician and Engineering Applications, 70*(2), 1641–1650.

Taigman, Y., Polyak, A., & Wolf, L. (2016). Unsupervised cross-domain image generation. arXiv: 1611.02200.

Tao, R., & Xu, J. (2023). Mapping with ChatGPT. *ISPRS International Journal of Geo-Information, 12*(7), 284.

Ullah, W., Zhang, Z., & Stefanidis, K. (2023). Sentiment analysis of mobile apps using BERT. *International Conference on Industrial, Engineering and Other Applications of Applied Intelligent Systems* (pp. 66–78). Springer.

Wang, F. Y., Miao, Q., Li, L., Ni, Q., Li, X., Li, J., Fan, L., Tian, Y., & Han, Q. L. (2024). When does Sora show: The beginning of TAO to imaginative intelligence and scenarios engineering. *IEEE/CAA Journal of Automatica Sinica, 11*(4), 809–815.

Weidinger, L., Uesato, J., Rauh, M., Griffin, C., Huang, P.S., Mellor, J., Glaese, A., Cheng, M., Balle, B., Kasirzadeh, A., & Biles, C. (2022). Taxonomy of risks posed by language models. In *Proceedings of the 2022 ACM Conference on Fairness, Accountability, and Transparency* (pp. 214–229).

Wiedemann, G., Remus, S., Chawla, A., & Biemann, C. (2019). Does BERT make any sense? Interpretable word sense disambiguation with contextualized embeddings. arXiv: 1909.10430.

Wu, T., He, S., Liu, J., Sun, S., Liu, K., Han, Q. L., & Tang, Y. (2023). A brief overview of ChatGPT: The history, status quo and potential future development. *IEEE/CAA Journal of Automatica Sinica, 10*(5), 1122–1136.

Yang, J., Jia, H., & Liu, H. (2022). Spatial relationship extraction of geographic entities based on Bert model. In *Journal of Physics: Conference Series* (Vol. 2363, No. 1, p. 012031). IOP Publishing.

Zhang, Y., Wei, C., He, Z., & Yu, W. (2024). GeoGPT: An assistant for understanding and processing geospatial tasks. *International Journal of Applied Earth Observation and Geoinformation, 131*, 103976.

Zhang, H., Xu, T., Li, H., Zhang, S., Wang, X., Huang, X., & Metaxas, D. N. (2017). Stackgan. Text to photo-realistic image synthesis with stacked generative adversarial networks. In *Proceedings of the IEEE International Conference on Computer Vision* (pp. 5907–5915), Venice, Italy, 22–29 October 2017.

Zhang, L., Qiu, Q., Lin, H., Zhang, Q., Shi, C., Yang, W., Shi, Y., Yang, S.; Xu, L., & Yu, J. (2023). DreamFace: Progressive Generation of Animatable 3D Faces under Text Guidance. arXiv:2304. 03117.

Zhang, Z., Amiri, H., Liu, Z., Züfle, A., & Zhao, L. (2023a). Large language models for spatial trajectory patterns mining. arxiv:2310.04942.

Zhao, B., Zhang, S., Xu, C., Sun, Y., & Deng, C. (2021). Deep fake geography? When geospatial data encounter Artificial Intelligence. *Cartography and Geographic Information Science, 48*(4), 338–352.

Zhou, E., & Lee, D. (2024). Generative artificial intelligence, human creativity, and art. *PNAS Nexus, 3*(3), pgae052.

Zhou, K. Z., Choudhry, A., Gumusel, E., & Sanfilippo, M. R. (2024). "Sora is Incredible and Scary": Emerging Governance Challenges of Text-to-Vide o Generative AI Models. arXiv:2406.11859.

Zhu, Z., Lu, J., Yuan, S., He, Y., Zheng, F., Jiang, H., Yan, Y., & Sun, Q. (2024). Automated generation and analysis of molecular images using generative artificial intelligence models. *The Journal of Physical Chemistry Letters, 15*(7), 1985–1992.

Chapter 4
Spatial AI in Robotics and Spatial Computing

Abstract The third distinctive facet of Spatial AI consists in types of AI that are embedded in devices dispersed in indoor or outdoor environments (i.e. in the form of Spatial Robotics and Ambient Intelligence), or at the intersections of AI with spatial computing (Augmented Reality, Metaverse, Digital Twins). The former may entail devices from nanochips to robots and the latter may involve spatially immersive forms of AI. More importantly, the contribution of AI in the development of Digital Twins for geospatial applications paves the way for the emulation of large scale spatial processes that may also entail AI-enabled holograms.

Keywords Spatial AI · Spatial computing · Augmented reality · Metaverse · Digital twins · AI-Holograms · Spatial robotics · SLAM · Ambient intelligence · Robotic ecologies

4.1 Spatial Robotics and Ambient Intelligence

AI can be a spatially distributed (embedded) technology, reflecting the underlying intersections of AI with spatial computing, hologram technologies, ambient intelligence and robotics. Augmented Reality (AR), the Metaverse, AI-enabled holograms and Ambient Intelligence, all add spatiality to AI, contributing to the creation of Spatial AI (Fig. 4.1).

The concept of "intelligent cities" (Komninos, 2011; Ratti & Claudel, 2016) entails a number of spatially-enabled technologies that render the physical space of a city "intelligent" and they can be as diverse as blockchain (Rejeb et al., 2022) and geosensor networks (Lee et al., 2012; Nittel et al., 2004) and as innovative as "Generative Urban Design Models". In these, AI is used to model streets, buildings, energy consumption, even the aesthetics of city design (Jiang et al., 2023). As fast as spatialized AI permeates the fabric of an intelligent city, GeoAI becomes increasingly important in recording and analyzing previously unnoticed patterns of society, enabling Spatial AI to encroach on "social sensing" (Aggarwal, 2013; Duan et al., 2020; Hu, 2018). Eventually, intelligent cities become endowed with their own

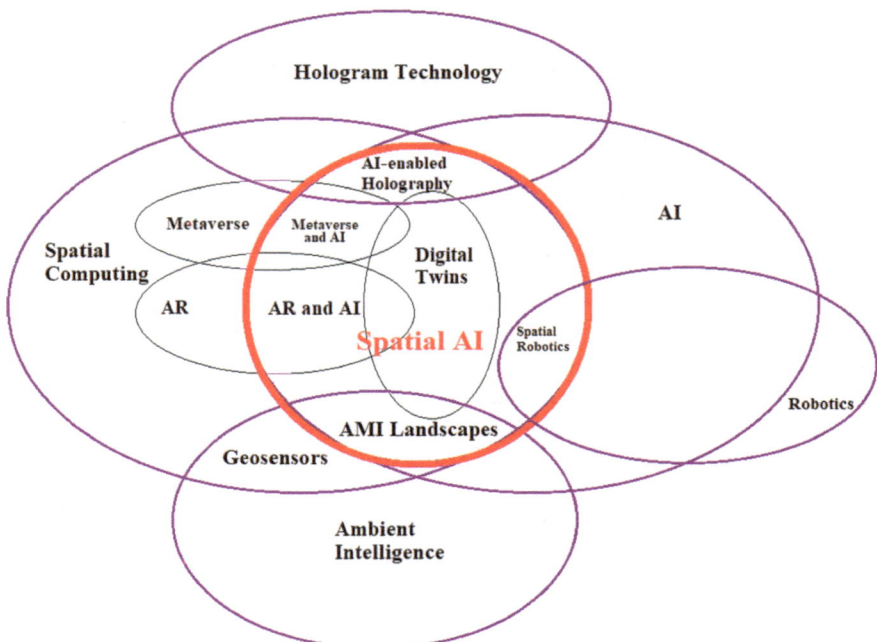

Fig. 4.1 Spatial AI lies at the heart of the intersections of AI with Spatial Robotics and Spatial Computing

"cyber-ecology" (Dyens, 1994, p. 327) that encompasses the spatial distribution and spatial interconnectedness of networks of devices, cables and wireless environments, in allocations and interactions reminiscent of natural ecosystems. In fact it has been anticipated that "interacting with humans, cyber physical systems will connect human society to form a new world—cyber physical society" (Shi & Zhuge, 2011, p. 972).

Besides cyber-ecology, "hybrid ecologies" and "ambient ecologies" (Crabtree & Rodden, 2008) highlight the interplay between mixed reality with ubiquitous computing environments. With an emphasis on the vision of ambient intelligence, ambient ecologies consist of smart objects, autonomous artifacts, portable devices, mobile robots, plain objects, services and people. Ambient ecologies "reside in smart environments" (Kameas & Saffiotti, 2012, p. 483) and the aims of ambient intelligence can be fulfilled in tandem with those of ambient ecology by integrating computational intelligence with smart objects, artificial agents, sensors and interaction interfaces (Surie et al., 2012); "smart homes" is a widely known examples of "Ambient Ecologies" (Mozer, 2005).

All these are forms of a "spatialized" AI, demonstrating the *materialization of Spatial AI* by entailing medical monitoring, traffic control, environmental management systems, navigation systems, emergency systems, robot navigation, entertainment, games and even interior decoration and art. In fact, ambient intelligence

(AmI) can easily be combined with AI, by using sensor chips, wireless data tranceivers, piezoelectric transmitters, often integrated with thermogenerators, flexible solar cells and adhoc wireless sensor and actuator networks etc., aiming to create physical objects that are able to perceive and analyze their own condition, and then proceed to changes if they calculate that they failed to meet their preset standards. These technological developments have led to the concept of "ambient intelligence landscapes", in which sensors (i.e. in the form of "smart dust") are scattered (systematically or randomly) in indoor environments or over physical landscapes. These types of *spatialized intelligence* present a vision for future intelligent spaces, where humans will be surrounded by integrated, responsive and personalised electronic environments, sensitive to human actions, movements and needs, emphasizing user-friendliness and access to information, anywhere and anytime. In these ways, ambient intelligence and ubiquitous computing increasingly relate to geoinformation technologies and the other way round.

Besides geo-sensors however, cyber-physical landscapes may involve robotic systems organized as "robotic ecologies" (Young Sung et al., 2010), exemplifying the coexistence of robots with humans and operating in hybrid human-and-robotic interaction environments. Robotic ecologies can be based on ALife-type simulations (Yang et al., 2019) and, interestingly, concepts of ethology may serve as guiding principles for designing them (Wilde & Murphy, 2019). In such simulations, the ecological models of natural ecosystems turn out to be powerful enough to aid in the exploration of robotic ecosystems (Egerstedt et al., 2018). The latter are designed to function as "autonomous agents", able to adopt to the spatial domains that are allocated to them (equivalent to the ecological "niche"). In "cognitive robotic ecologies" (Saffiotti & Broxvall, 2005) in particular, "future work should explore scenarios in which users and cognitive robotic ecologies collaborate and learn to exploit their different capabilities to their mutual benefit" (Dragone et al., 2015, p. 279), which is akin to "adaptive ecologies" that involve autonomous humanoid robots (Rahman, 2013). The field of robotic ecologies thus intersects with that of evolutionary robotics, which combines various tracks of ALife, swarm intelligence and cyber-ecologies (Bongard, 2013; Nolfi & Floreano, 2000).

Researching robotic adaptation to spatial domains broadens the scope of Artificial Human Intelligence (AHI) in relation to adaptation to spatial settings of outdoor or indoor environments. The problem of SLAM (simultaneous localization and mapping) is very characteristic of the links between Spatial AI and robotics. SLAM is encountered when robots are used for mapping small or large geographical settings (Alsadik & Karam, 2021; Bailey & Durrant-Whyte, 2006; Blochliger et al., 2018; Chang et al., 2007; Hong et al., 2021) and it is a double spatial problem: mapping out the immediate surroundings (by sensors if by a robot or by a mental map if by a human) and simultaneously locating oneself in it. While humans solve SLAM problems in fractions of a second, devising robots to do it fast has turned out to be a complex problem that (to this date) has been mainly tackled by using Bayesian methods for updating spatial information about the environment, involving sensors and LIDAR arrays.

Further, the importance of spatial analysis in the formation of robotic ecologies becomes evident from the fact that topology is crucial in setting up wireless networks (Santi, 2005), whereas the allocation of sensors follows topologies that determine the performance of sensor networks (de Silva & Ghrist, 2007). Topology determines the movements of robotic UAVs and, as a matter of fact, lattice topologies (triangular, square, ring topologies etc.) have a central role in optimizing robotic performance in the geographical space (Huo et al., 2020) since they emerge in Spatial AI in arranging mobile robotic networks, as well in the creation of robot swarms, e.g. networked robots acting in a coordinated fashion for monitoring large areas (Qi et al., 2015).

4.2 AI in Spatial Computing

The introduction of AI in Augmented Reality technologies (AR) elevates the spatialization of AI higher by connecting physical space, virtual space and AI together (Bassyouni & Elhajj, 2021) and creating forms and applications of AI that can be immersed in AR environments (Fig. 4.2).

The Metaverse, a virtual 3D world, was first used as a term in 1992 in Neal Stephenson's novel "Snow Crash". Before its recent adoption by Facebook's creator, the term "Metaverse" was used in parallel with virtual worlds such as "Second Life" and various video games. The popular virtual world "Second Life" for instance, is a

Fig. 4.2 A surface created by applying an Ant Optimization Algorithm inserted in an AR immersive environment

vast "Metaverse" in which spatial arrangements resemble the real world (land parcels, regions, private islands, microcontinents, mainland etc.), while also allowing for social interactions through avatars. Once such a private virtual world has been created, a wide range of opportunities emerges for producing beautiful artistic creations (i.e. gardens filled with digital entities with striking resemblance to real plants). The relationships between AI and the Metaverse have been explored in the context of diverse domains: robotics, Extended Reality, Internet of Things, Generative Artificial Intelligence Digital Twins whereas Spatial AI is instrumental for the development of the Metaverse (Soliman et al., 2024; Thakur et al., 2023). And, despite the current applications of AI in the Metaverse being hitherto confined to education and healthcare, more and powerful intersections with IoT and XR are anticipated in the future (Jauhiainen, 2024; Liu & Siau, 2023; Lv, 2023).

Simultaneously, and in efforts to bypass the use of not always user-friendly VR devices, the field of AI-enabled holograms has been gaining ground during the last years. This technology applies AI methods to create spatial 3D holograms with AR that may pop up from the user's screen (i.e. from a smartphone). Expectedly, AI-enabled holograms constitute a disruptive technology with a very wide scope of applications (Al Shaghroud et al., 2023; Patel & Bhalodiya, 2019; Situ, 2022; Zhong, 2024), particularly in gaming, education and medicine. The Australian government for instance, has launched holographic technologies offering immersive experiences that allow users to visualize different alternative scenarios, by combining AR with AI (Allam & Jones, 2021).

The range of "Spatial Computing" encompasses the technology of Digital Twins as well. Digital Twins use advanced analytics and machine intelligence in order to produce reliable predictions about the physical entity they relate to. The foremost important characteristic of a Digital Twin is its "mirroring" the physical world, presenting a digital replica of the physical entity that, beyond models and simulations, also communicates with the physical entity it refers to, in a bilateral communication. In geospatial applications, a Digital Twin is expected to simulate processes of spatial dynamics as faithfully as possible and then deliver information and suggestions i.e. for land management. So long as the interactions between a Digital Twin and the physical object it purportedly simulates remain unhampered, a Digital Twin is expected to simulate a physical system with real-time synchronization of the data emitted from the physical system to the Digital Twin and the reverse. This process may eventually be used to optimize the structure and function of the physical system, according to the recommendations that the Digital Twin comes up with (Fig. 4.3).

Although the impact of Digital Twins is significantly higher if designated to simulate urban environments, their usefulness is not confined to such geographical settings only, since they can also be used to optimize land management of rural areas too (Pylianidis et al., 2021; Verbouw et al., 2021; Zhai et al., 2020), i.e. in smart farming, in planning agricultural landscapes, or even in planning for planetary-scale ecosystem management (DeFelipe et al., 2022). Digital Twins have also been proposed (both as concept and tehnology) for land use management of smart cities (Akroyd et al., 2022), for combining geospatial technologies with social media and volunteered geographic information as well as for geographical applications

Fig. 4.3 A Digital Twin is expected to reflect the patterns and processes of spatial dynamics of the physical world and then deliver suggestions for spatial monitoring, analysis and management

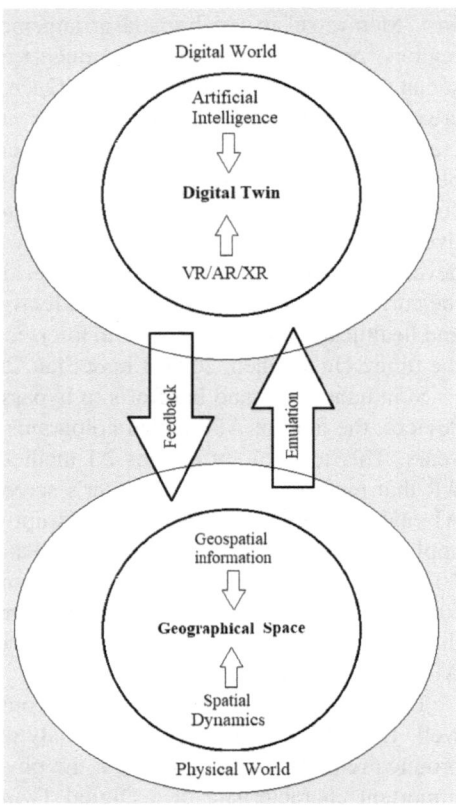

involving VR and AR (Luo et al., 2022). Besides, one of the far-reaching impacts of Digital Twins for spatial analysis consists in the possibility they offer for the contribution of public geospatial information to decision-making (Abdeen & Sepasgozar, 2021). Whether they are used for urban or rural geographical areas or for both, Digital Twins must satisfy some conditions to properly fulfill their purpose: firstly, they must integrate and interface all the necessary data from a physical region so as to adequately mirror its characteristics; secondly, all the information they handle must be effectuated in real time and thirdly, they must be useful for geographical predictions.

The gradual spatialization of AI through Digital Twins becomes evident from the fact that AI has manifold applications in Digital Twin technologies (Emmert-Streib, 2023) in the context of applied geography, environmental management, sustainability and resilience assessments, in optimizing transportation infrastructure (Wu et al., 2022) and in other domains. Besides, AI can be embedded in Digital Twins also through other technologies, such as cyber-physical systems (Radanliev et al., 2022), VR/AR, robotics (Huang et al. 2021), Internet of Things (Zhang et al. 2022; Zayed et al., 2023) and the Metaverse (Bordegoni & Ferrise, 2023).

The wide range of spatially-enabled applications (e.g. related to floods, climate change) embedded in the range of functionalities of Digital Twins is reflected by two projects of the European Union: "Destination Earth" (Nativi et al., 2020) and "Gaia-X: A Federated Infrastructure for Europe" (EU, 2021), in which is mentioned the concept of "emergent behaviors" of digital ecosystems and their co-evolution with physical ecosystems in order to maximize resilience at planetary scale (Nativi et al., 2020, p. 9). Indeed, the aim of "Destination Earth" is to achieve a full replica of the earth by 2030, so the following passage is perhaps characteristic of the E.U. project "Destination Earth" (from the website https://digital-strategy.ec.europa.eu/en/policies/destination-earth):

> What is Digital Twin Earth? The European Union is finalizing plans for an ambitious "digital twin" of planet Earth that would simulate the atmosphere, ocean, ice, and land with unrivaled precision, providing forecasts of floods, droughts, and fires from days to years in advance Destination Earth (DestinE) and its development of digital earth twins are key to predicting the effects and building resilience to climate-change. Destination Earth (DestinE) aims to develop – on a global scale - a highly accurate digital model of the Earth to monitor and predict the interaction between natural phenomena and human activities.

Digital Twins are not only expected to be able to model and simulate spatial dynamics with high accuracy, rapidity and reliability, but also to "emulate" spatial dynamics. Emulation requires setting up convincing models and simulations, so much so that the emulated entity can hardly be distinguished from the original. Without the aid of some fairly advanced AI however, the vision of emulation seems unfeasible, since the leap from simulation/modeling to emulation can not be made without making major breakthroughs in handling spatial complexity. In fact, it presupposes that particular mathematical models will have been adopted to describe and forecast spatial changes and that alternative plans for spatial planning and/or management will have been incorporated, so that a Digital Twin becomes capable of simulating the feedback that the physical spatial system would feed it with. Yet, in the context of current technologies, models and applications, an elaborate Digital Twin for geographical applications might be able to develop AI for spatial monitoring, assessing, forecasting and decision making based on different scenarios of long-term land use changes that can be based on existing deterministic (Papadimitriou, 2023a, b) or stochastic models (Papadimitriou, 2023c). Further still, the integration of AI with a Digital Twin may be designed to yield predictions of qualitative changes in spatial behaviors such as resilience, self-organization and emergence (Papadimitriou, 2022a, b, 2023d, e, f).

References

Abdeen, F. N., & Sepasgozar, S. M. E. (2021). City digital twin concepts: A vision for community participation. *Environmental Sciences Proceedings, 12*, 19.

Aggarwal, C. C. (Ed.). (2013). *Managing and mining sensor data.* Springer.

Akroyd, J., Harper, Z., Soutar, D., Farazi, F., Bhave, A., Mosbach, S., & Kraft, M. (2022). Universal digital twin: Land use. *Data-Centric Engineering, 3,* e3.

Al Shaghroud, S., Al Shuwaier, A., & Al Rakaf, L. (2023). Artificially intelligent and interactive 3D hologram. *International Conference on Human-Computer Interaction* (pp. 367–373). Springer.

Allam, Z., & Jones, D. S. (2021). Future (post-COVID) digital, smart and sustainable cities in the wake of 6G: Digital twins, immersive realities and new urban economies. *Land Use Policy, 101,* 105201.

Alsadik, B., & Karam, S. (2021). The simultaneous localization and mapping (SLAM)-an overview. *Journal of Applied Science and Technology Trends, 2*(02), 147–158.

Bailey, T., & Durrant-Whyte, H. (2006). Simultaneous localization and mapping (SLAM): Part II. *IEEE Robotics & Automation Magazine, 13*(3), 108–117.

Bassyouni, Z., & Elhajj, I. H. (2021). Augmented reality meets artificial intelligence in robotics: A systematic review. *Frontiers in Robotics and AI, 8,* 724798.

Blochliger, F., Fehr, M., Dymczyk, M., Schneider, T., & Siegwart, R. (2018). Topomap: Topological mapping and navigation based on visual slam maps. In *2018 IEEE International Conference on Robotics and Automation (ICRA)* (pp. 3818–3825). IEEE.

Bongard, J. (2013). Evolutionary robotics. *Communications of the ACM, 56*(8), 74–83.

Bordegoni, M., & Ferrise, F. (2023). Exploring the intersection of metaverse, digital twins, and artificial intelligence in training and maintenance. *Journal of Computing and Information Science in Engineering, 23*(6), 060806.

Chang, H. J., Lee, C. G., Hu, Y. C., & Lu, Y. H. (2007). Multi-robot SLAM with topological/metric maps. In *2007 IEEE/RSJ international conference on intelligent robots and systems* (pp. 1467–1472). IEEE.

Crabtree, A., & Rodden, T. (2008). Hybrid ecologies: Understanding cooperative interaction in emerging physical-digital environments. *Personal and Ubiquitous Computing, 12,* 481–493.

De Silva, V., & Ghrist, R. (2007). Coverage in sensor networks via persistent homology. *Algebraic and Geometric Topology, 7,* 339–358.

DeFelipe, I., Alcalde, J., Baykiev, E., Bernal, I., Boonma, K., Carbonell, R., Flude, S., Folch, A., Fullea, J., García-Castellanos, D., & Geyer, A. (2022). Towards a digital twin of the earth system: Geo-soft-CoRe, a geoscientific software and code repository. *Frontiers in Earth Science,* 509.

Dragone, M., Amato, G., Bacciu, D., Chessa, S., Coleman, S., DiRocco, M., Gallicchio, C., Gennaro, C., Lozano, H., Maguire, L., McGinnity, M., Micheli, A., O'Hare, G. M. P., Renteria, A., Saffioti, A., Vairo, C., & Vance, P. (2015). A cognitive robotic ecology approach to self-configuring and evolving AAL systems. *Engineering Applications of Artificial Intelligence, 45,* 269–280.

Duan, W., Chiang, Y. Y., Leyk, S., Uhl, J. H., & Knoblock, C. A. (2020). Automatic alignment of contemporary vector data and georeferenced historical maps using reinforcement learning. *International Journal of Geographical Information Science, 34*(4), 824–849.

Dyens, O. (1994). The emotion of cyberspace: Art and cyber-ecology. *Leonardo, 27*(4), 327–333.

E.U. (2021). GAIA-X Consortium. GAIA_X: A Federated Data Infrastructure for Europe, April 2021. Available at: https://www.datainfrastructure.eu/GAIAX/Navigation/EN/Home/home.html

Egerstedt, M., Pauli, J. N., Notomista, G., & Hutchinson, S. (2018). Robot Ecology: Constraint-based control design for long duration autonomy. *Annual Reviews in Control, 46,* 1–7.

Emmert-Streib, F. (2023). What is the role of AI for digital twins? *AI, 4*(3), 721–728.

Hong, S., Bangunharcana, A., Park, J. M., Choi, M., & Shin, H. S. (2021). Visual SLAM-based robotic mapping method for planetary construction. *Sensors, 21*(22), 7715.

Hu, Y. (2018). Geo-text data and data-driven geospatial semantics. *Geography Compass, 12*(11), e12404.

Huang, Z., Shen, Y., Li, J., Fey, M., & Brecher, C. (2021). A survey on AI-driven digital twins in industry 4.0: Smart manufacturing and advanced robotics. *Sensors, 21*(19), 6340.

Huo, X., Yang S., Lian B., Sun T., & Song Y.(2020). A survey of mathematical tools in topology and performance integrated modeling and design of robotic mechanism. *Chinese Journal of Mechanical Engineering, 33*, 62.

Jauhiainen, J. S. (2024). The Metaverse: Innovations and generative AI. *International Journal of Innovation Studies, 8*(3), 262–272.

Jiang, F., Ma, J., Webster, C. J., Chiaradia, A. J., Zhou, Y., Zhao, Z., & Zhang, X. (2023). Generative urban design: A systematic review on problem formulation, design generation, and decision-making. *Progress in Planning*, 100795.

Kameas, A. & Saffiotti, A. (2012). Editorial. Special Issue on "Ambient Ecologies". *Pervasive and Mobile Computing, 8*, 483–484.

Komninos, N. (2011). Intelligent cities: Variable geometries of spatial intelligence. *Intelligent Buildings International, 3*(3), 172–188.

Lee, Y., Jung, Y. J., Nam, K. W., Nittel, S., Beard, K., & Ryu, K. H. (2012). Geosensor data representation using layered slope grids. *Sensors, 12*(12), 17074–17093.

Liu, Y., & Siau, K. L. (2023). Generative Artificial Intelligence and Metaverse: Future of Work, Future of Society, and Future of Humanity. In *International Conference on AI-generated Content* (pp. 118–127). Springer Nature.

Luo, J., Liu, P., & Cao, L. (2022). Coupling a physical replica with a digital twin: A comparison of participatory decision-making methods in an urban park environment. *ISPRS International Journal of Geo-Information, 11*(8), 452.

Lv, Z. (2023). Generative artificial intelligence in the metaverse era. *Cognitive Robotics, 3*, 208–217.

Mozer, M. C. (2005). Lessons from an adaptive Home. In D. J. Cook & S. K. Das (Eds.), *Smart environments: technology, protocols, and applications* (pp. 273–298). Wiley.

Nativi, S., Delipetrev, B., & Craglia, M. (2020). Destination earth: Survey on "Digital Twins" technologies and activities, in the Green Deal area, EUR 30438 EN, Publications Office of the European Union, Luxembourg, ISBN 978-92-76-25160-6. https://doi.org/10.2760/430025, JRC122457

Nittel, S., Stefanidis, A., Cruz, I., Egenhofer, M., Goldin, D., Howard, A., Labrinidis, A., Madden, S., Voisard, A., & Worboys, M. (2004). Report from the first workshop on geo sensor networks. *ACM SIGMOD Record, 33*(1), 141–144.

Nolfi, S., & Floreano, D. (2000). *Evolutionary robotics: The biology, intelligence, and technology of self-organizing machines.* MIT Press.

Papadimitriou, F. (2022a). *Spatial entropy and landscape analysis.* Springer VS.

Papadimitriou, F. (2022b). Emergence, sustainability and cyber-physical landscapes. *Spatial entropy and landscape analysis* (pp. 123–139). Springer VS.

Papadimitriou, F. (2022c). Spatial negentropy and social self-organization in simulated landscapes. *Spatial entropy and landscape analysis* (pp. 75–86). Springer VS.

Papadimitriou, F. (2023a). *Modelling landscape dynamics: Determinism, stochasticity and complexity.* Springer VS.

Papadimitriou, F. (2023b). Dynamical systems modelling of landscape transformations. *Modelling landscape dynamics: determinism, stochasticity and complexity* (pp. 1–15). Springer VS.

Papadimitriou, F. (2023c). Modelling nonlinear and complex landscape dynamics. *Modelling landscape dynamics: Determinism, stochasticity and complexity* (pp. 17–26). Springer VS.

Papadimitriou, F. (2023d). Markov models of landscape dynamics. *Modelling landscape dynamics: determinism, stochasticity and complexity* (pp. 45–57). Springer VS.

Papadimitriou, F. (2023e). Modelling landscape resilience. *Modelling landscape dynamics: determinism, stochasticity and complexity* (pp. 101–118). Springer VS.

Papadimitriou, F. (2023f). Complexity, non-locality and riddledness in landscape dynamics. *Modelling landscape dynamics: determinism, stochasticity and complexity* (pp. 119–133). Springer VS.

Patel, D., & Bhalodiya, P. (2019). 3D holographic and interactive artificial intelligence system. In *2019 International conference on smart systems and inventive technology (ICSSIT)* (pp. 657–662). IEEE.

Pylianidis, C., Osinga, S., & Athanasiadis, I. N. (2021). Introducing digital twins to agriculture. *Computers and Electronics in Agriculture, 184*, 105942.

Qi, Y., Sun, T., Song, Y., & Jin, Y. (2015). Topology synthesis of three-legged spherical parallel manipulators employing Lie group theory. *Proceedings of the Institution of Mechanical Engineers, Part c: Journal of Mechanical Engineering Science, 229*(10), 1873–1886.

Radanliev, P., De Roure, D., Nicolescu, R., Huth, M., & Santos, O. (2022). Digital twins: Artificial intelligence and the IoT cyber-physical systems in Industry 4.0. *International Journal of Intelligent Robotics and Applications, 6*(1), 171–185.

Rahman, S. M. M. (2013). People-centric adaptive social ecology between intelligent autonomous humanoid robot and virtual human for social cooperation. *Communications in Computer and Information Science, 413*, 120–125.

Ratti, C., & Claudel, M. (2016). *The city of tomorrow: Sensors, networks, hackers, and the future of urban life.* Yale University Press.

Rejeb, A., Rejeb, K., Simske, S. J., & Keogh, J. G. (2022). Blockchain technology in the smart city: A bibliometric review. *Quality & Quantity, 56*(5), 2875–2906.

Saffiotti, A. & Broxvall, M. (2005). PEIS Ecologies: Ambient Intelligence meets autonomous robotics. In: Proceedings of the 2005 joint conference on smart objects and ambient intelligence: innovative context-aware services: usages and technologies. ACM,, pp. 277–281.

Santi, P. (2005). *Topology control in wireless Ad Hoc and sensor networks.* Wiley.

Shi, X., & Zhuge, H. (2011). Cyber physical socio ecology. *Concurrency and Computation: Practice and Experience, 23*(9), 972–984.

Situ, G. (2022). Deep holography. *Light Advanced Manufacturing, 3*(2), 278–300.

Soliman, M. M., Ahmed, E., Darwish, A., & Hassanien, A. E. (2024). Artificial intelligence powered metaverse: Analysis, challenges and future perspectives. *Artificial Intelligence Review, 57*(2), 36.

Surie, D., Janlert, L.-E., Pederson, T., & Roy, D. (2012). Egocentric interaction as a tool for designing ambient ecologies—The case of the easy ADL ecology. *Pervasive and Mobile Computing, 8*, 597–613.

Thakur, S. S., Bandyopadhyay, S., & Datta, D. (2023). Artificial intelligence and the metaverse: present and future aspects. *The Future of Metaverse in the Virtual Era and Physical World* (pp. 169–184). Springer.

Verbouw, C., Tekinerdogan, B., Beulens, A., & Wolfert, S. (2021). Digital twins in smart farming. *Agricultural Systems, 189*, 103046.

Wilde, G. A., & Murphy, R. (2019). A robotics-oriented taxonomy of how ethologists characterize the traversability of animal environments. *Robotics and Autonomous Systems, 118*, 159–166.

Wu, J., Wang, X., Dang, Y., & Lv, Z. (2022). Digital twins and artificial intelligence in transportation infrastructure: Classification, application, and future research directions. *Computers and Electrical Engineering, 101*, 107983.

Yang, H., Cao, S., Bai, L., Zhang, Z., & Kong, J. (2019). A distributed and parallel self-assembly approach for swarm robotics. *Robotics and Autonomous Systems, 118*, 80–92.

Young Sung, J., Grinter, R. E., & Christensen, H. I. (2010). Domestic robot ecology. an initial framework to unpack long-term acceptance of robots at home. *International Journal of Social Robotics, 2*(4), 417–429.

Zayed, S. M., Attiya, G. M., El-Sayed, A., & Hemdan, E. E. D. (2023). A review study on digital twins with artificial intelligence and internet of things: Concepts, opportunities, challenges, tools and future scope. *Multimedia Tools and Applications, 82*(30), 47081–47107.

Zhai, Z., Martínez, J. F., Beltran, V., & Martínez, N. L. (2020). Decision support systems for agriculture 4.0: Survey and challenges. *Computers and Electronics in Agriculture, 170*, 105256.

Zhang, Z., Wen, F., Sun, Z., Guo, X., He, T., & Lee, C. (2022). Artificial intelligence-enabled sensing technologies in the 5G/internet of things era: From virtual reality/augmented reality to the digital twin. *Advanced Intelligent Systems, 4*(7), 2100228.

Zhong, W. (2024). Application of artificial intelligence digital holography technology based on medical sensors in the development of medical image fusion. *Measurement: Sensors, 33*, 101146.

Chapter 5
Spatial AI for Artificial General Intelligence

Abstract Certain capabilities of Spatial AI may indicate that steps towards the development of spatially-enabled Artificial General Intelligence (AGI) may have already been made, thus prompting for the formulation of criteria that would enable us to decide whether a Spatial AI has attained the status of AGI. Towards this aim, a set of ten criteria is presented here. Specifically, a spatially-enabled AGI should be able to: calculate basic metrics related to spatial objects, associate 2D and 3D shapes and patterns with semantics as humans do, recognize spatial entities (even if rotated, inverted or viewed from different angles), identify hierarchies of shapes and objects, give correct directions for navigation (and solve SLAM problems), advise on spatial decision making and spatial planning, make simple predictions of future spatial distributions, associate spatial forms with dynamics and motion, display evidence of creativity of spatial forms that can be aesthetically pleasing.

Keywords Spatial AI · Artificial general intelligence (AGI) · AI creativity · AI criteria · AGI criteria · Spatial AGI

5.1 Spatial Criteria for AGI

Janowicz et al., (2020, p. 631) posed a key question concerning the future of GeoAI: "Can we develop an artificial GIS analyst that passes a domain-specific Turing Test by 2030?". Interestingly, it was shown that a ChatGPT can pass a GIS exam test with GPT-4, scoring as high as 88.3% (Mooney et al., 2023). In the near future, Spatial AI may be enriched with new neural networks models such as "liquid neural networks" (Chahine et al., 2023; Hasani et al., 2021) and "spiking neural networks" that mimic human brain activity by means of "spike" thresholds; a concept to be found also behind another form of neural networks, the "neuromorphic" NNs (Ghosh-Dastidar & Adeli, 2009; Yamazaki et al., 2022). Such NNs may contribute towards the development of Artificial General Intelligence (AGI), endowing Spatial AI with more flexibility, versatility and adjustability to new inputs (which may nevertheless be at the expanse of explainability) and enabling AI to reason like a human. Beyond AGI however, Artificial Super Intelligence (ASI) is considered to be the ultimate stage of development of AI (Bostrom, 2014), clearly surpassing both human intelligence and AGI.

Spatial AGI is expected to be able to perform spatial tasks, analyses and data processing with the speed, efficacy and accuracy that humans do. A set of ten criteria (by no means a final or restrictive list) for Spatial AGI is given here. So an AI that has attained the status of AGI should be able to:

- Calculate basic metrics related to spatial objects in euclidean space (i.e. distance and area measurements);
- Associate 2D and 3D shapes and patterns with semantics as humans do (e.g. associate the word "chair" to the shape of a chair);
- Recognize such shapes, even if they are rotated, inverted, or viewed from different angles;
- Identify hierarchies of 2D and 3D shapes and objects of the real world as humans do (e.g. if it's a tree, then be able to identify what particular species it is and "understand" that the legs of a chair are constituents of the object "chair");
- Give correct directions for navigation (upon being fed with appropriate maps) and solve SLAM problems as rapidly as humans are able to;
- Identify simple topological relationships among spatial objects in 3D space from texts and images (i.e. from sentences of the type "the cup is on the table");
- Advise on spatial decision making and spatial planning (provided that it is fed with appropriate decision diagrams);
- Make simple predictions of future spatial distributions of points, areas, populations etc. spatial entities on the basis of predefined models, with spatial and temporal accuracy no higher than what would be expected from a human;
- Associate spatial forms with dynamics and motion, where appropriate;
- Display evidence of creativity of spatial forms that can be aesthetically pleasing.

Further to the above, a spatially-enabled AGI should be able to handle spatial uncertainties that may emerge in the course of time (evidently these are different than semantic uncertainties such as those that GeoAI already deals with). For instance, a spatially-enabled AGI should be able to identify equivalences among spatial objects that are isomorphic to one another as much as humans are, i.e. interpreting correctly how rotation, inversion, translation and other elementary geometric transformations may alter the appearance of a map and thus be capable of classifying a set of square maps (Fig. 5.1) into different equivalence classes:

```
Class I: a, r, s, t
Class II: b, e, f, g, h, i, j
Class III: c, d, o, p
Class IV: k, l, m, n
```

Yet, some deeper insight into "Spatial AGI" can be gained by considering the following problem: two maps (Fig. 5.2) with the same Shannon entropy (H_1) can be quite different visually, with approximately 50% different spatial complexities

Fig. 5.1 A set of square maps with 3 colors each

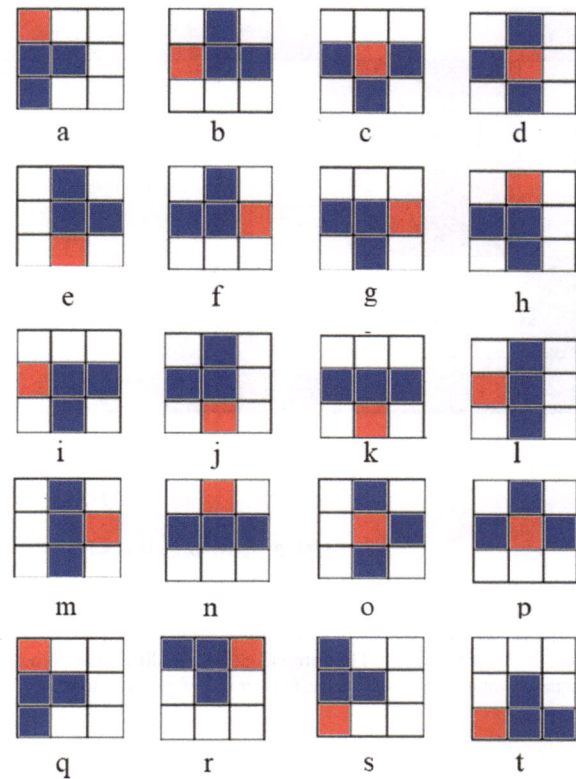

measured by means of the spatial complexity measures C_{P1} and C_{P2} (Papadimitriou 2020a, 2020b), but they nevertheless may obey a common rule governing the spatial allocation of colors (four white cells per column and an equal number per row as well). Discovering the hidden regularity in these two maps may be a task fairly easy to accomplish for a human or for AGI, provided that the problem has been posed in a suitable way. For instance, asking AGI to:

```
Find what similarity is there between these two maps
```

it will probably quickly succeed to identify the similarity in counts of white cells. But if confronted with the reverse task:

```
Devise two binary 8x8 maps,
both with the same spatial entropy,
but with different spatial complexities
```

then this might be considerably more demanding for a human and probably as much for an AGI, possibly encroaching onto the range of capabilities expected from ASI.

AGI should also be able to apply Jordan's theorem to surfaces and thus automatically decide the topological situation of a point with respect to a closed curve (whether it is inside or outside), depending on the number of intersection points that

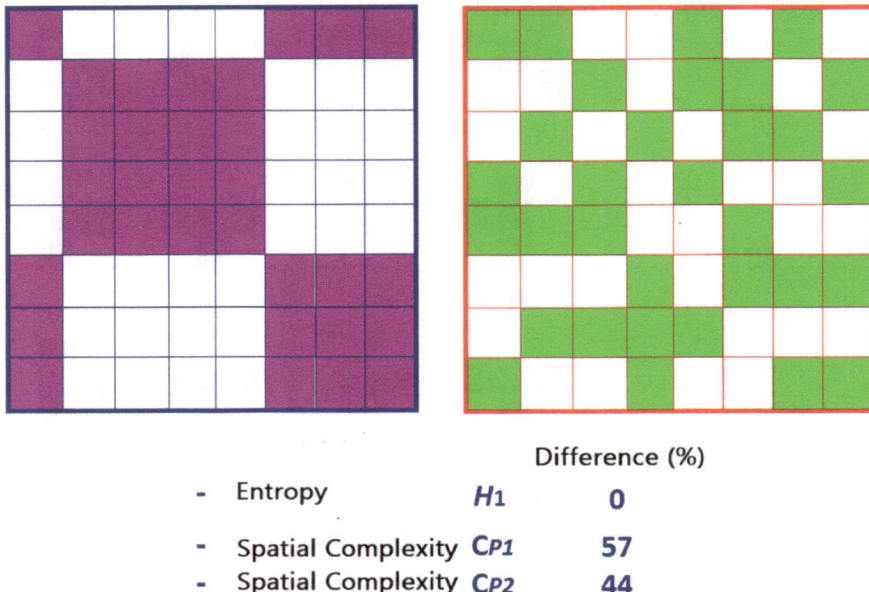

			Difference (%)
-	Entropy	H_1	0
-	Spatial Complexity	C_{P1}	57
-	Spatial Complexity	C_{P2}	44

Fig. 5.2 Two maps with the same Shannon entropy (H_1) and with their spatial complexities differing by 50% approximately. They are visually quite different but they both obey a simple geometric rule: equal numbers of colored cells (4) per row and per column

any rays emanating from that point will have with the boundary of the closed curve (Fig. 5.3). Yet, it is rather beyond the capabilities of AGI (as is for humans') to unfurl a complex 2D shape to simplify its topology so as to apply Jordan's theorem on it (Fig. 5.4).

Aside of these, as spatial entities change in the course of time, it is interesting to explore the extent to which AGI might be able to forecast and interpret spatial dynamics. Let us consider a simple spatial simulation that can be created using an ABM with three land uses (F = forest, A = agriculture and S = shrubland) interacting in 5-cell Von Neumann neighborhoods (Fig. 5.5). The basic rule here is that if at least two cells of the same land use type are found within the same Von Neumann neighborhood, then that land use type prevails. If two land use types are represented by two cells each, then ecological criteria prevail that are employed to decide which land use type will result in each particular situation. Hence, all three land use types interact according to an overall decision plan (Table 5.1) that applies to each and all cells of the map, at Von Neumann neighborhoods.

By applying these rules, it can be seen that this ABM yields oscillatory behaviors: the initial map allocation at step 1 leads to an alternation of allocations at time steps 3 and 4 and ad infinitum after the 5th time step (Fig. 5.6) and hence the model has no decisive final spatial allocation.

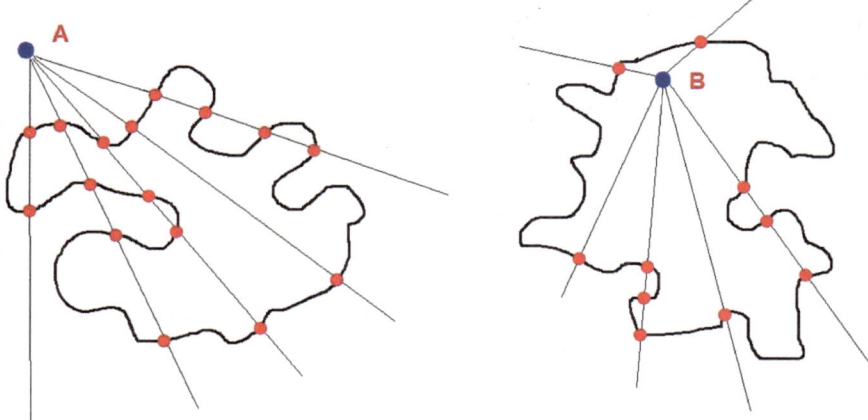

Fig. 5.3 An even number of intersection points indicates that a point is outside a closed curve, while an odd number of intersection points indicates it is found inside the closed curve

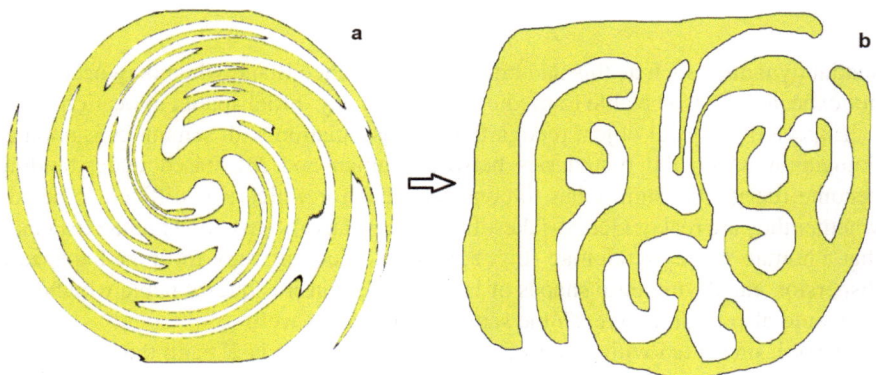

Fig. 5.4 AGI is not expected to unfurl a 2D shape so as to simplify its topology and infer which part of the plane is inside a closed curve

This **ABM** encapsulates the phenomenon of spatial synchrony, which has been observed in natural ecosystems in many instances (Williams & Liebhold, 2000; Engen & Sæther, 2016; Peltonen et al., 2002) and may arise from changes in population sizes or spatial interactions that cause synchronic reactions of different populations to some external periodic event (Liebhold et al., 2004). As in this case, it may emerge from simple rules, whereas one of the most interesting facets of applying Von Neumann neighborhoods to spatial analysis is their offering the possibility to study uncertainty (Meng et al., 2015; Qin et al., 2017; Cheng & Zheng, 2019; Shang et al., 2022) or oscillating behaviors (Binder & Jaramillo, 1997; Kennedy & Mendes, 2006) in spatial domains. While AGI can be trained to identify peculiar behaviors in

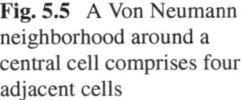

Fig. 5.5 A Von Neumann neighborhood around a central cell comprises four adjacent cells

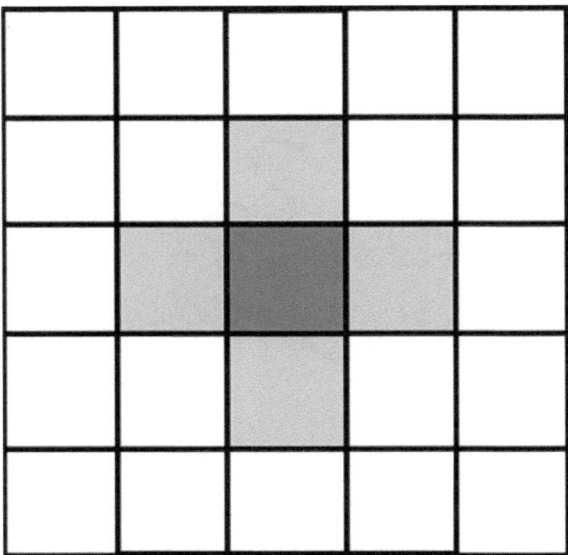

spatial dynamics such as this and affirm their presence if and when they occur, it is not expected to also be able to predict them from any initial spatial allocation.

Expectedly, spatial object recognition and association with semantics is another domain in which the boundaries between human and ML-based understanding become fuzzy. One simply has to consider maps for which it is difficult to decide whether they are real or GenAI-fakes. It may be relatively easy for a human to decide that a human-made pseudomap (Fig. 5.7) is fake, judging only from the haphazard dispersion and asymmetric shapes of land parcels, but it is questionable whether an AI would also be able to reach the same conclusion as well.

Also, if presented with a pair of images of which one is 2D and the other is 3D (and with the same color palette), then AGI should be able to establish that the two images are closely related (Fig. 5.8).

5.2   AGI and Spatial Creativity

In what concerns AI creativity, Spatial AI is currently already capable of yielding creative artistic production (paintings, to this day) and this creativity is character-ized by plasticity, vitality and diversity (Liu, 2020), although AI creativity is still inadequate to replace human subjectivity in aesthetic judgment (Braga & Logan, 2017). However, new avenues open up for visual aesthetics (Mazzone & Elgammal, 2019; Qi, 2019), although it is very likely that the interpretations of AI creations will present significant divergences. Yet, some psychological or philosophical theories

Table 5.1 Decision plan for the various combinations of F = forest, A = agriculture and S = shrubland at a Von Neumann neighborhood around a central cell and the resulting cell in each case

Central	Von Neumann Neighborhood			Result	Central	Von Neumann Neighborhood			Result	Central	Von Neumann Neighborhood			Result
	F	A	S			F	A	S			F	A	S	
F	4	0	0	F	A	4	0	0	F	S	4	0	0	F
	3	1	0	F		3	1	0	F		3	1	0	F
	2	2	0	F		2	2	0	F		2	2	0	F
	2	1	1	F		2	1	1	F		2	1	1	F
	2	0	2	S		2	0	2	S		2	0	2	S
	1	1	2	S		1	1	2	S		1	1	2	S
	1	2	1	A		1	2	1	A		1	2	1	A
	0	2	2	A		0	2	2	A		0	2	2	A
	0	3	1	A		0	3	1	A		0	3	1	A
	0	1	3	S		0	1	3	S		0	1	3	S
	0	4	0	A		0	4	0	A		0	4	0	A
	0	0	4	S		0	0	4	S		0	0	4	S
	1	0	3	S		1	0	3	S		1	0	3	S
	1	3	0	A		1	3	0	A		1	3	0	A
	3	0	1	F		3	0	1	F		3	0	1	F

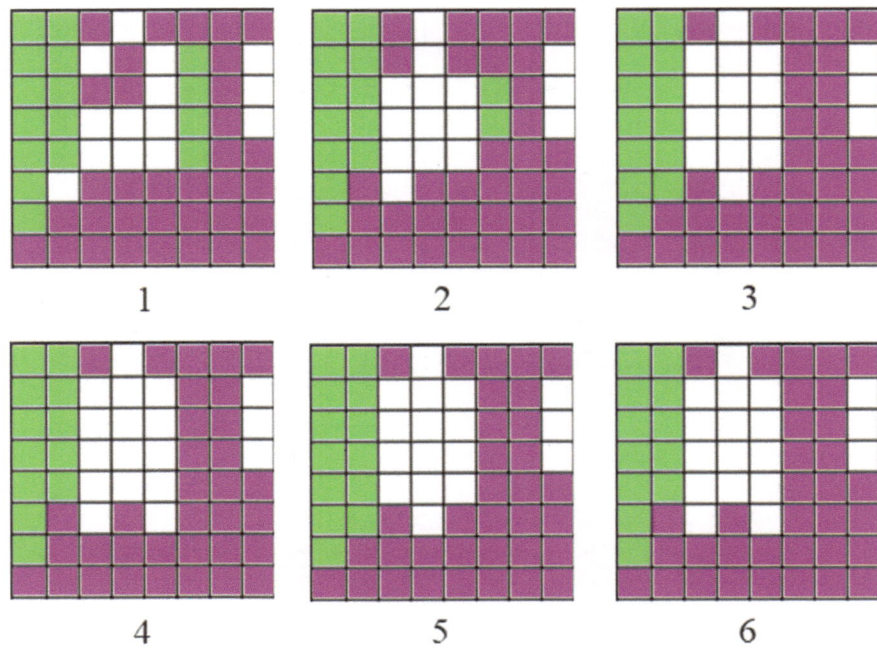

Fig. 5.6 The evolution of an ABM model in time steps 1 to 6

Fig. 5.7 An edge-enhanced image of a meticulously made-up (fake) map

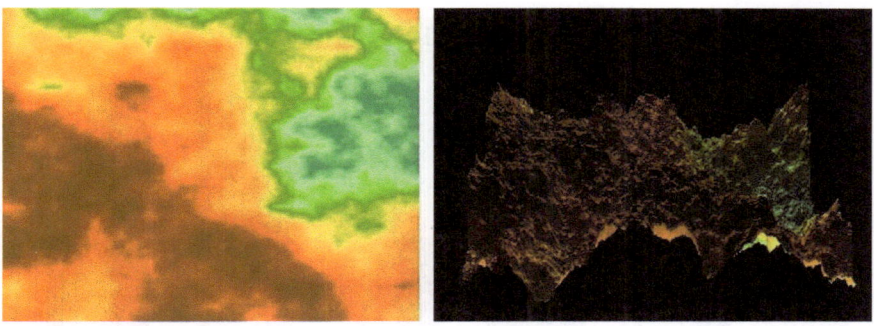

Fig. 5.8 An AGI should be able to establish the relevance or similarity between 2 and 3D imagery

of creativity and aesthetics might be applicable to both human-made and GenAI-made creations. One such is "honing theory" that highlights the "potentiality" of alternative possible creative outcomes by "superposition" and has been adopted to explain the creations produced by DeepDream (Di Paola et al., 2018). In practice, it is often difficult to discern human-made from AI-made paintings, since fake paintings created by convolutional neural networks may already display aesthetically pleasing features (such as color harmony and resemblance to known artistic currents) and with the same variety and intensity as human-made paintings. In fact, it may not be far from the truth to assert that the spectacular performance of AI in the visual arts testifies that Spatial AI has already achieved a status equal to (if not higher than) that of AGI. In this respect, quantitative analyses linking image aesthetics with spatial complexity and spatial entropy (Papadimitriou, 2020a, 2020b, 2022a, 2022b, 2022c, 2024a, 2024b) might be explored deeper in order to identify possible relationships among objectively measured spatial indices of images and their aesthetic appeal.

Technically, the results of imagery produced by GenAI are most commonly evaluated by means of specifically-designed metrics that take into account the spatial characteristics of the image output. Hence, the hitherto available metrics aim at evaluating the quality of images and their similarity. As for quality, the most commonly used measures are accuracy and realism score, while usual metrics for similarity assessment are specific structural similarity indexes and Fréchet inception distances. But human judgement may also be a precious (if not the final and most important) evaluator.

Expectedly, the human perspective also matters quite as much as objective (quantative) methods in evaluating such imagery. For this reason, a research project that was organized by the author was carried out aiming to assess the extent by which University students of the University of West Attica as well as graduates of GIS, geology and geographical education ($n = 30$) were able to distinguish photographs of landscapes shot by humans from GenAI-made fakes. Eight photographs of real landscapes were randomly interchanged with as many images created by AI and the

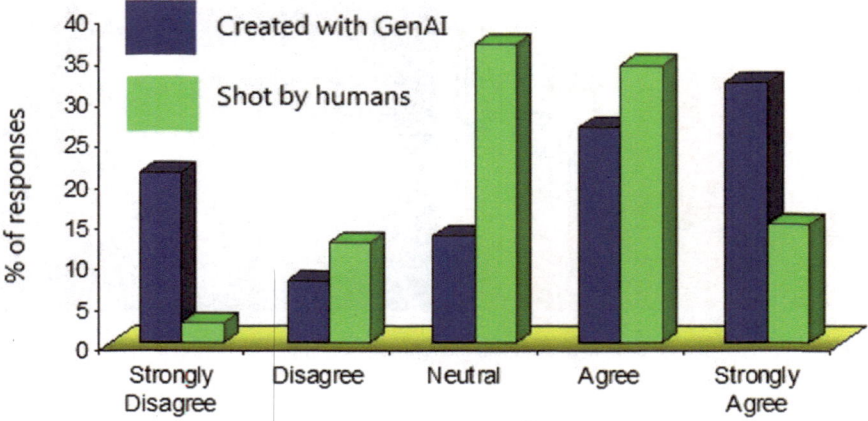

Fig. 5.9 The responses of the participants on whether a photograph depicts a real landscape or an AI-fake show that the imagery created with AI can be convincing enough so as to confuse the viewer whether the image depicted a real geographical setting or not

set of 16 images was presented to the subjects. The photographs of the real land-scapes were from Greece, Brazil, Thailand and Austria. The AI-fakes were created with the aid of various programs that are freely available online. The responses to the simple prompt: "This picture was created by AI" were framed to the standard 5-grades Likert scale (Fig. 5.9). The differences between agreements and disagreements turned out to be statistically significant for both AI fakes and real landscapes ($\chi^2 = 16.2724$, P-value $= 0.000055$; significant at $p < 0.01$), while there was no statistically significant difference in the responses between students and graduates ($\chi^2 = 1.719$, P-value $= 0.18982$). The percentages of responses demonstrating confidence that the images were created by AI were about as high as those for real landscapes. Yet, more than 25% disagreed that the AI-fakes were indeed created by AI, implying that some AI-fakes can be convincing enough so as to confuse the viewers whether the image depicted some real geographical setting or not.

References

Binder, P. M., & Jaramillo, J. F. (1997). Stabilization of coherent oscillations in spatially extended dynamical systems. *Physical Review E, 56*(2), 2276.

Bostrom, N. (2014). *Superintelligence: Paths, dangers, strategies.* Oxford University Press.

Braga, A., & Logan, R. K. (2017). The emperor of strong AI has no clothes: Limits to artificial intelligence. *Information, 8*(4), 156.

Chahine, M., Hasani, R., Kao, P., Ray, A., Shubert, R., Lechner, M., Amini, A., & Rus, D. (2023). Robust flight navigation out of distribution with liquid neural networks. *Science Robotics, 8*(77), eadc8892.

Cheng, Y., & Zheng, X. (2019). Effect of uncertainty on cooperative behaviors during an emergency evacuation. *Communications in Nonlinear Science and Numerical Simulation, 66*, 216–225.

DiPaola, S., Gabora, L., & McCaig, G. (2018). Informing artificial intelligence generative techniques using cognitive theories of human creativity. *Procedia Computer Science, 145*, 158–168.

Engen, S., & Sæther, B. E. (2016). Spatial synchrony in population dynamics: The effects of demographic stochasticity and density regulation with a spatial scale. *Mathematical Biosciences, 274*, 17–24.

Ghosh-Dastidar, S., & Adeli, H. (2009). Spiking neural networks. *International Journal of Neural Systems, 19*(04), 295–308.

Hasani, R., Lechner, M., Amini, A., Rus, D., & Grosu, R. (2021). Liquid time-constant networks. *Proceedings of the AAAI Conference on Artificial Intelligence, 35*(9), 7657–7666.

Janowicz, K., Gao, S., McKenzie, G., Hu, Y., & Bhaduri, B. (2020). GeoAI: Spatially explicit artificial intelligence techniques for geographic knowledge discovery and beyond. *International Journal of Geographical Information Science, 34*(4), 625–636.

Kennedy, J., & Mendes, R. (2006). Neighborhood topologies in fully informed and best-of-neighborhood particle swarms. *IEEE Transactions on Systems, Man, and Cybernetics, Part C (Applications and Reviews), 36*(4), 515–519.

Liebhold, A., Koenig, W. D., & Bjørnstad, O. N. (2004). Spatial synchrony in population dynamics. *Annual Review of Ecology Evolution and Systematics, 35*, 467–490.

Liu, X. (2020). Artistic reflection on artificial intelligence digital painting. *Journal of Physics: Conference Series, 1648*(3), 032125.

Mazzone, M., & Elgammal, A. (2019). Art, creativity, and the potential of artificial intelligence. *Arts, 8*(1), 26.

Meng, X. K., Xia, C. Y., Gao, Z. K., Wang, L., & Sun, S. W. (2015). Spatial prisoner's dilemma games with increasing neighborhood size and individual diversity on two interdependent lattices. *Physics Letters A, 379*(8), 767–773.

Mooney, P., Cui, W., Guan, B., & Juhász, L. (2023). Towards understanding the geospatial skills of ChatGPT: Taking a geographic information systems (GIS) exam. In *Proceedings of the 6th ACM SIGSPATIAL International Workshop on AI for Geographic Knowledge Discovery* (pp. 85–94).

Papadimitriou, F. (2020a). *Spatial Complexity. Theory, Mathematical Methods and Applications.* Springer.

Papadimitriou, F. (2020b). Spatial complexity, visual complexity and aesthetics. *Spatial complexity: Theory, mathematical methods and applications* (pp. 243–261). Springer.

Papadimitriou, F. (2020c). The algorithmic basis of spatial complexity. *Spatial complexity: Theory, mathematical methods and applications* (pp. 81–99). Springer.

Papadimitriou, F. (2022a). *Spatial entropy and landscape analysis.* Springer VS.

Papadimitriou, F. (2022b). Visual perception of spatial entropy and landscape analysis. *Spatial entropy and landscape analysis* (pp. 87–102). Springer VS.

Papadimitriou, F. (2022c). Spatial entropy, geo-information and spatial surprise. *Spatial entropy and landscape analysis* (pp. 1–14). Springer VS.

Papadimitriou, F. (2024a). *Geo-topology: Theory, models and applications.* Springer.

Papadimitriou, F. (2024b). Geo-topology and Visual Impact. *Geo-topology: theory, models and applications* (pp. 139–150). Springer.

Peltonen, M., Liebhold, A. M., Bjørnstad, O. N., & Williams, D. W. (2002). Spatial synchrony in forest insect outbreaks: Roles of regional stochasticity and dispersal. *Ecology, 83*(11), 3120–3129.

Qi, W. (2019). Research on Art of Artificial Intelligence from the Perspective of Symbolic Aesthetics. *3rd international conference on art studies: science, experience, education (ICASSEE 2019)* (pp. 642–644). Atlantis Press.

Qin, J., Chen, Y., Fu, W., Kang, Y., & Perc, M. M. (2017). Neighborhood diversity promotes cooperation in social dilemmas. *IEEE Access, 6*, 5003–5009.

Williams, D. W., & Liebhold, A. M. (2000). Spatial synchrony of spruce budworm outbreaks in eastern North America. *Ecology, 81*(10), 2753–2766.

Yamazaki, K., Vo-Ho, V. K., Bulsara, D., & Le, N. (2022). Spiking neural networks and their applications: A review. *Brain Sciences, 12*(7), 863.

Chapter 6
Spatial AI for Artificial Super-Intelligence

Abstract This chapter explores how the spatial capabilities of Artificial Superintelligence (ASI) would differ from those of AGI. After examining various digital representations of spatial entities (maps, photographs, diagrams etc.) as indicative examples, it is possible to delineate boundaries and similarities among human intelligence, AGI and ASI. It is suggested that ASI is expected to be able to perform advanced geometric, statistical and topological assessments and analyses of spatial objects, distinguish photographs of physical objects from computer simulations, identify 3D shapes, infer spatial dynamics from shapes, associate meanings with spatial entities and associate spatial allocations with non-spatial information. Further, ASI should be able to *generate* spatial forms in ways that neither humans nor AGI could possibly do, i.e. by integrating algorithms that have been developed in earlier stages of development of AI.

Keywords Spatial AI · Artificial superintelligence (ASI) · Superintelligence · Artificial general intelligence (AGI) · Spatial complexity · AI criteria

6.1 Spatial Indicators for ASI

Artificial Superintelligence (ASI) is the ultimate stage of evolution of AI (Yampolskiy, 2015; Gill, 2016; Sotala, 2017; Baum et al., 2017; Narain et al., 2019; Jabari and Lundborg, 2021 and (if or when it is realized), it is anticipated to clearly surpass the capabilities of AGI (Bostrom, 2014).

While a human readily understands that some figure is not a photograph but a computer artifact instead (Fig. 6.1), an AGI should be able to confirm this as well (although it may not be able to identify exactly *how* an artifact might have been created), while an ASI is expected to be able to tackle this problem *if* the spatial entity was created by some known models or set of algorithms. Plausibly however, one of the foremost important spatial capabilities of an ASI is expected to be its beyond-human capability of image interpretation and spatial pattern recognition. For instance, ASI is expected to be able to tell whether an image portrays a real

© The Author(s), under exclusive license to Springer Nature Switzerland AG 2025   65
F. Papadimitriou, *Spatial Artificial Intelligence*,
SpringerBriefs in Computational Intelligence,
https://doi.org/10.1007/978-3-031-82136-3_6

spatial object or a computer creation when neither a human nor AGI might do so (Figs. 6.2 and 6.3).

Further, spatially-enabled ASI should be endowed with the capacity to outperform humans (and AGI) in the spatial computing of 3D shapes. For instance, it should exceed humans and current AI tools in its capability to calculate, in reasonable time, the vertices, faces, edges, holes and cavities of a complex voxelized 3D domain from the statistical frequencies of occurrences of geometric and topological characteristics at the outer surface (or surfels) of the spatial object (Fig. 6.4); a computational task that neither a human nor an AGI could possibly carry out in reasonable time and with acceptable confidence.

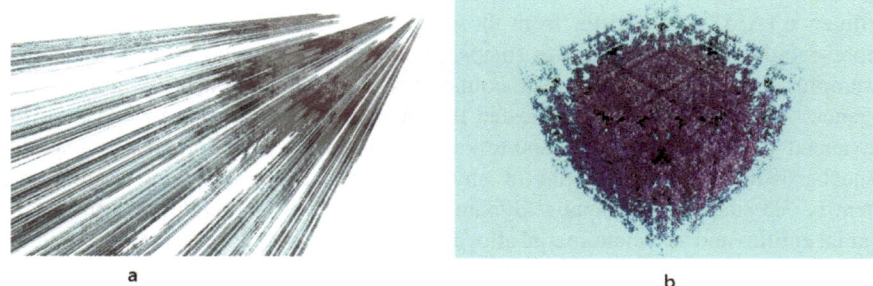

a   b

Fig. 6.1 Both humans and AGI would find it easy to decide that these shapes are indeed computer artifacts, although an ASI may go one step further by also confirming that they are a kaleidoscopic IFS fractal (**a**) and a Mandelbox (**b**)

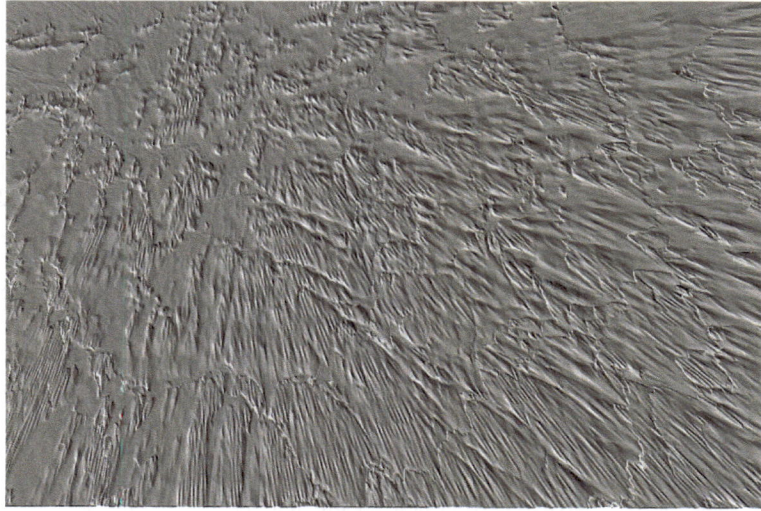

Fig. 6.2 Both humans and AGI would find it hard to decide whether this image portrays a real object (i.e. an extraterrestrial planetary surface) or a computer-made surface

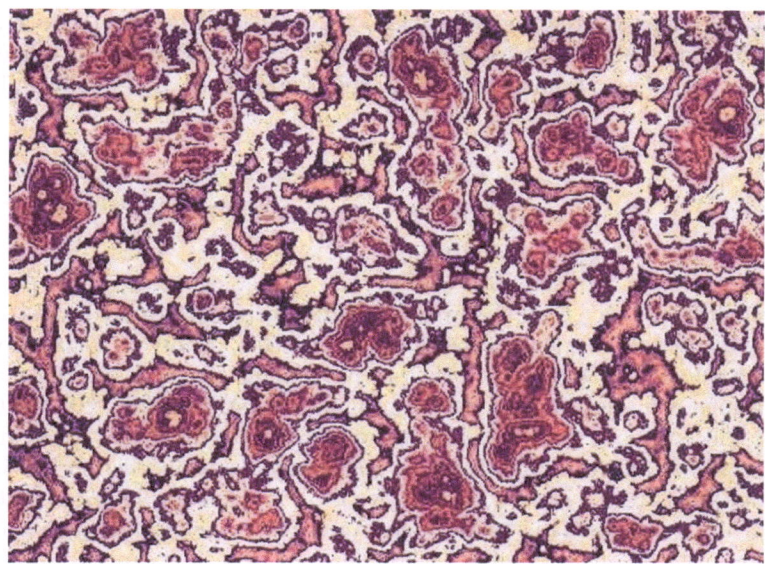

Fig. 6.3 An ASI should be able to tell whether this is a snapshot of a histopathological examination or a computer-generated fake

Fig. 6.4 A voxelized 3D spatial domain

Very much like AGI is expected to able to affirm the identity of two cellular maps that have been rotated, translated or inverted, ASI is expected to be able to identify 3D shapes even if seen from a perspective angle. For instance, ASI would be able to affirm that a knot, i.e. created by the set of parametric equations

$$x = 0.56\cos(0.87\theta) + 1.1\cos(-3.67\theta)$$
$$y = 0.56\sin(0.87\theta) + 1.1\sin(-3.67\theta)$$
$$z = 0.45\sin(0.37\theta)$$

is indeed the same knot, even if viewed from different perspectives. But, it is highly unlikely that it will be equally capable of asserting the identity of a spatial object with its rotation-symmetric if the object is overly complex (Fig. 6.5). This may be an indication that spatial complexity (Papadimitriou, 2020a, 2020b, 2020c, 2024a, 2024b) will emerge as the prevailing and decisive factor in spatial object recognition by either AGI or ASI, whether a spatial object is embedded in 2D, in 3D (Papadimitriou, 2020a), or even 4D or higher dimensions (Papadimitriou, 2020b).

Even if viewed statically (without motion), images giving the impression of motion are instantly interpreted as such by humans, so ASI is expected to be able to also identify the underlying spatial dynamics that has produced the shapes and colors that describe a moving object. Although AGI is expected to interpret an image as displaying some kind of motion or movement as a human would, it is an open problem whether ASI might cope effectively with what might be called a "semantic deception". Consider, i.e. an image that does *not* display dynamics but it only looks as if it did; one such (Fig. 6.6) is produced from the graphic plot based on the trigonometric equation

$$f(x, y) = \tan^{-1}\left(\tan\left(\sin(xy)\sin\left(\frac{xy}{\pi}\right) + x + \sin x + y\right) + y\right)$$

which is not a function of time and yet, it may easily give the (false) impression to either humans or AI that it is a snapshot of motion.

And yet, the high expectations of an ASI should not be confined to object recognition only, for they should also extend to the interpretation of meanings associated with spatial domains.

Consider a multi-colored map (Fig. 6.7) with (13×13)-1 $= 168$ cells (the bottom right cell is excluded) with spatial complexities (Papadimitriou, 2020b) equal to C_{P1} $= 168$ (in relative terms compared to the map's maximum possible spatial complexity $C_{P1} = 168/168 = 100\%$) and $C_{P2} = 135$ (in relative terms $135/144 = 93.75\%$).

Such high spatial complexity values (93.75% and 100% of the complexity that would be expected from complete spatial randomness) diminish the possibility of establishing any association of this image with anything meaningful. But if the cells were made to correspond to letters of the greek alphabet, then a run length encoding reveals the message: a kind of a (distasteful) dish described by the ancient Greek comic play writer Aristophanes (446-386 bC) within a single word ("Εκκλησιάζουσαι" Ekklesiazousai, verse 1163):

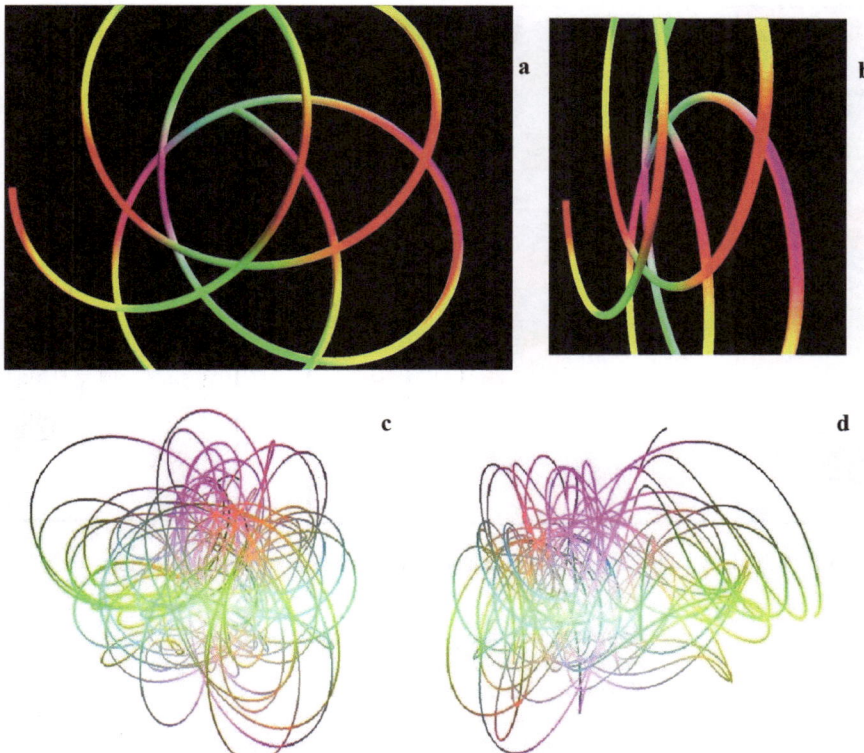

Fig. 6.5 An ASI might be able to decide whether two perspective views (**a** and **b**) are of the same knot. Yet, if a 3D spatial object is overly complex, then even ASI is not expected to be able to assert whether two perspective views are indeed different views of the same object (as they are here) or not (**c** and **d**)

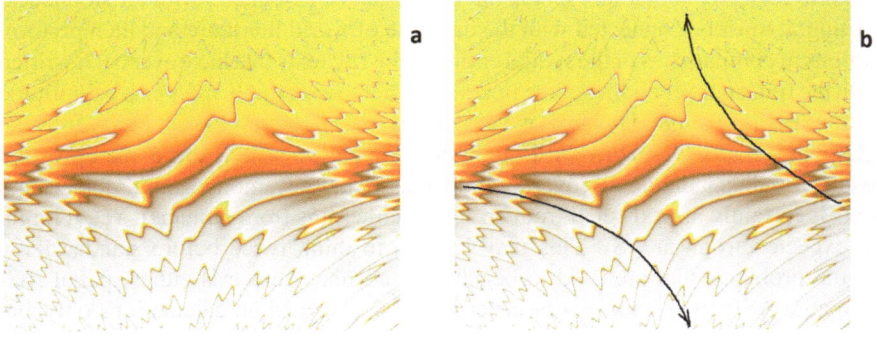

Fig. 6.6 Both humans and AI can misinterpret shapes and colors of an image (**a**) as depicting some motion or movement (**b**) while, in fact, it does not

Fig. 6.7 A spatially highly complex map that corresponds to a word from a greek text

Λοπαδοτεμαχοσελαχογαλεοκρανιολειψανο
Διμυποτριμματοσιλφιοκαραβομελιτο
Κατακεχυμενοκιχλεπικοσσυφοφαττο
Περιστεραλεκρυονοπτοκεφαλλιοκιγκλο
πελειολαγωσιραιοβαφητραγανοπτερύγων

This example shows how meanings can correspond to (or be embedded in) spatial entities and our spatial analysis tools may still be inadequate to reveal them. Truly, a Spatial AI that would easily reveal meanings from spatial allocations of symbols, classes of objects and, eventually, numbers, will be extraordinarily advanced. For an AI to understand that this spatial allocation can be associated with Aristophanes' writing, it must be connected with the database of world literature and then perform zillions of comparisons of the spatial entity against all texts that have ever been written with the run length encoding of the image, in order to eventually infer exact similitude with a text quotation. This task is certainly well beyond the expectations from an AGI, but a spatially-enabled ASI might perform such a search and come up with a result. By all standards, discovering hidden associations among spatial objects and texts is certainly not expected from an AGI (neither it is from humans). Yet, it might be expected from an ASI. This may lead us to think twice before standardizing spatial problems that are used to benchmark advances towards either spatial AGI or spatial ASI: such standards or criteria may not depend on calculations only, but on interpretations of spatial allocations as well (by linking them with non-spatial information).

This image may also give us the chance to gain some deeper insight: all its cells encoded to correspond with the letter alpha (α) are located at the following 19 positions of the run length encoding:

4, 10, 16, 20, 26, 34, 48, 58, 60, 70, 72, 96, 108, 123, 141, 147, 151, 156, 158.

Now consider a random sample of a number of $k = 5$ cells that are labeled as alpha and located at the following positions:

$$10, 48, 96, 123, 141.$$

Then, an AGI should be able to provide an estimate of the image size from these five numbers by making use of the simple rough approximation formula that calculates the length of a series of numbers only from the maximum number sampled (max) and the sample size k:

$$N = \max +(\max -k)/k$$

In this case, the maximum of the sampled numbers is 141 and thus:

$$N = 141 + (141 - 5)/5 = 141 + 27.2 = 168.2$$

which is fairly close to the true image size (168).

But, if more symbol series were fed to the AI with different k values each, then an ASI should be able to provide significantly more accurate estimates of the image size (or even derive statistical estimates) in a way that neither a human nor an AGI might.

6.1.1 Creativity in Spatial ASI

Aside of shape, object, pattern and movement recognition, ASI should be able to *generate* spatial forms in ways that neither humans not AGI could possibly do. CNNs, DNNs or other NNs of the future will certainly be indispensable in the creation of a generative ASI. Yet, besides NNs, such creative processes might exploit a whole universe of already known algorithms that are available from software that has been created during the last decades.

Simulating plant leaves for instance, Prusinkiewicz and Lindenmeyer (1990), developed the first artificial (digital) floral forms based on Lindenmeyer's algorithms for "L-systems", by using "parallel graph languages" that generate computer-made simulations of forms of trees, leaves, flowers, herbs, bushes. These computer-made simulations were not merely graphically designed copies of real plants; they were algorithmically developed plant forms, leaf by leaf, stem by stem, with algorithmic rules mimicking the growth and branching of plants in nature. So following the same set of algorithmic rules iteratively, it is possible to simulate plant forms. One characteristic example of such algorithmic methods is the "Barnsley Fern". This is a fractal form similar to the leaves of the real fern black spleenwort-*Asplenium adamantum-nigrum* (Barnsley, 2014). Creating a procedurally generated form of a leaf begins from the origin (0,0) in 2d-space and more points are added as four functions (f_1 to f_4) are calculated at random, with f_1 used for 1% of the time, f_2 for 85%, f_3 for 7% and f_4 for 7% of the time:

$$f_1$$
$$x_{n+1} = 0$$
$$y_{n+1} = 0.16y_n$$
$$f_2$$
$$x_{n+1} = 0.85x_n + 0.04y_n$$
$$y_{n+1} = -0.04x_n + 0.85y_n + 1.6$$
$$f_3$$
$$x_{n+1} = 0.2x_n - 0.26y_n$$
$$y_{n+1} = 0.23x_n + 0.22y_n + 1.6$$
$$f_4$$
$$x_{n+1} = -0.15x_n + 0.28y_n$$
$$y_{n+1} = 0.26x_n + 0.24y_n + 0.44$$

Each one of these functions describes a different part of the plant: f_1 for the stem, f_2 for the small leaflets and f_3, f_4 for the largest left-hand and right-hand leaflets respectively. Consequently, these four functions are interrelated by means of the affine transformation:

$$f(x, y) = \begin{bmatrix} a & b \\ c & d \end{bmatrix} \begin{bmatrix} x \\ y \end{bmatrix} + \begin{bmatrix} e \\ f \end{bmatrix}$$

As the values of a, b, c, d, e, f change, the leaves of different plant species can be simulated: *Pellaea, Nephrolepis, Culcita dubia, Cyclosorus* or *Theltpteridaceae* fern etc. In these simulations, each one of the functions is allocated a different "probability" (that is equivalent to program running time).

The possibility to digitally reproduce a spatial setting by selecting appropriate algorithms is probably one of the most important characteristics of a generative spatial ASI and this goes well beyond planar figures such as plant leaves. For instance, ASI should be able to reproduce the image of a real landscape by means of an algorithm or an equation. Consider the (not very high admittedly) similarity between a real landscape photograph and the graphic representation of the equation (Fig. 6.8):

$$f(x, y) = \left[\frac{(0.3 - \tan^{-1} y) - \frac{\pi}{\log_2(9.7)}}{\pi y + x \sin y} \right] \left(\frac{(x - \sin y)(0.3 - \tan^{-1} y)}{\pi^2 y \log_2(9.7)} \right)$$

shows that, with current computational capacities, this is possible (even if only in the reverse way: knowing the equation's plot and matching it with a known picture).

Developing Spatial AI towards ASI may require new technologies, although already available AI methods can also be used. For instance, it might be possible to contribute to it by using agent-based models also (Ponomarev & Voronkov, 2017). Thus, linking the progress in CNNs (or their equivalent models) with known algorithms and with powerful computational routines that have already been developed in the context of available software will speed up the process towards creating spatial ASI.

Fig. 6.8 A real landscape (above, view from Pendeli mountain towards Evia, 25 km north of Athens, Greece) and a digital creation on the basis of an equation (below)

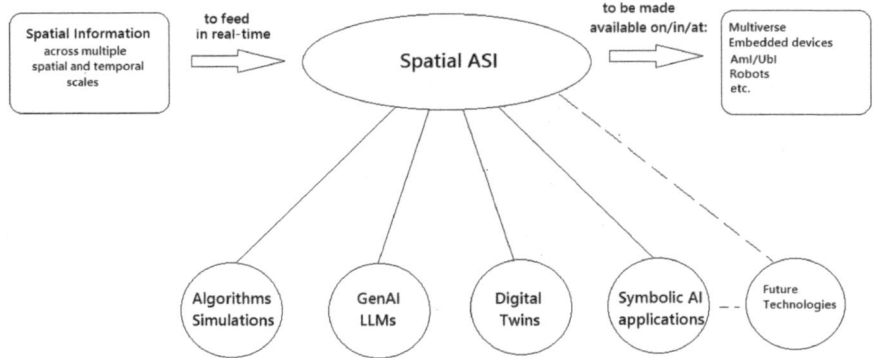

Fig. 6.9 Overview of central methods and technologies that can contribute towards the development of Spatial ASI

Additionally, any spatially-enabled software or know-how that has been developed in earlier stages of the development of AI (i.e. in the framework of Symbolic AI) may also be gradually incorporated into ASI, so as to endow it with expertise that has been developed by innumerable experts across the world. The same applies to software and/or datasets that Digital Twins rely on and hence, once an interconnected world-wide network of spatially-enabled Digital Twins becomes possible, it will significantly contribute to developing the spatial capacities of ASI (Fig. 6.9).

References

Barnsley, M. F. (2014). *Fractals everywhere*. Academic.

Baum, S., Barrett, A., & Yampolskiy, R. V. (2017). Modeling and interpreting expert disagreement about artificial superintelligence. *Informatica, 41*(7), 419–428.

Bostrom, N. (2014). *Superintelligence: Paths, dangers, strategies*. Oxford University Press.

Gill, K. S. (2016). Artificial super intelligence: Beyond rhetoric. *AI & Society, 31*, 137–143.

Jebari, K., & Lundborg, J. (2021). Artificial superintelligence and its limits: Why AlphaZero cannot become a general agent. *AI & Society, 36*(3), 807–815.

Narain, K., Swami, A., Srivastava, A., & Swami, S. (2019). Evolution and control of artificial super-intelligence (ASI): A management perspective. *Journal of Advances in Management Research, 16*(5), 698–714.

Papadimitriou, F. (2020a). *Spatial complexity. Theory, mathematical methods and applications*. Springer.

Papadimitriou, F. (2020b). Exploring spatial complexity in 3d. *Spatial complexity: Theory, mathematical methods and applications* (pp. 101–113). Springer.

Papadimitriou, F. (2020c). Spatial complexity in 4-and-higher-dimensional spaces. *Spatial complexity: Theory, mathematical methods and applications* (pp. 115–123). Springer.

Papadimitriou, F. (2024a). *Geo-topology: Theory, models and applications*. Springer Nature.

Papadimitriou, F. (2024b). Geo-topology and epistemology. *Geo-topology: theory, models and applications* (pp. 163–171). Springer.

Ponomarev, S., & Voronkov, A. E. (2017). Multi-agent systems and decentralized artificial superintelligence. arXiv preprint arXiv:1702.08529.

Prusinkiewicz, P., & Lindenmayer, A. (1990). *The algorithmic beauty of plants*. Springer.

Prusinkiewicz, P. (1998). In search of the right abstraction: The Synergy between Art, Science and Information Technology in the Modeling of Natural Phenomena. In C. Sommerer, & L. Mignonneau, (Eds.), *Art @ Science* (pp. 60–68). Springer-Verlag.

Sotala, K. (2017). How feasible is the rapid development of artificial superintelligence? *Physica Scripta, 92*(11), 113001.

Yampolskiy, R. V. (2015). *Artificial superintelligence: A futuristic approach*. CRC Press.

Chapter 7
Spatial AI, Big Data, Quantum AI and Transcomputation

Abstract Spatial AI is based on computation over spatial domains. Calculations suggest that, due to what can be named "the miracle of spatial information", the number of spatial questions required to identify the location of a square cell on it tends to zero as the size of a square map increases. Yet, physics sets an upper bound (10^{93} bits) to any kind of computing on the face of the earth (the Bremermann limit) and this limit applies irrespective of the stage of development of AI. So the expansion of big spatial data and spatial combinatorial explosions prompt us to examine possible limits to the capabilities of Spatial AI. While the Landauer limit might be overcome by Quantum AI, avoiding the obstacle of transcomputation in spatial data processing seems more challenging. However, as calculated here, classical computing implies that the total number of bits required to process all possible square binary map configurations does not exceed the size 17×17, while for maps 10×10 the calculation of all possible configurations will be possible by classic computation for only up to 8 colors.

Keywords Spatial AI · Quantum artificial intelligence · Transcomputation · Big geospatial data · Bremermann limit · Landauer limit · Spatial complexity · Limits to computation · Spatial computing

7.1 Quantum AI for Spatial AI

By all assessments, Quantum Artificial Intelligence (QAI) is expected to give a tremendous boost to classic AI, particularly in what concerns machine learning (Abdelgaber & Nikolopoulos, 2020; Ayoade et al., 2022; Moret-Bonillo, 2015; Zhu & Yu, 2023) since quantum computing is spectacularly faster than classical computing and capable of parallel processing of large volumes of data.

Quantum convolutional neural networks are created by using qubits (Henderson et al., 2020) so methods of quantum deep learning can be used to exploit the power

© The Author(s), under exclusive license to Springer Nature Switzerland AG 2025   77
F. Papadimitriou, *Spatial Artificial Intelligence*,
SpringerBriefs in Computational Intelligence,
https://doi.org/10.1007/978-3-031-82136-3_7

of generative adversarial networks (Garg & Ramakrishnan, 2020), while these techniques do not preclude hybrid approaches also, i.e. using classical-quantum algorithms (Rivas et al., 2021). Likewise in classical computing, for any quantum state, a quantum neural network also entails convolution and pooling, but the data in QAI are defined in quantum infintite-dimensional Hilbert space instead of the ordinary euclidean space.

One of the domains that QAI may contribute to Spatial AI is in the solution (under certain conditions) of some hitherto computationally hard problems of spatial optimization (Guo & Wang, 2020; Shokry & Youssef, 2021). Supervised, unsupervised and reinforcement learning may all benefit from quantum computing, particularly when big spatial data are to be dealt with, such as those derived from satellite imagery (Ayoade et al., 2023; Ghosh et al., 2024); NASA-QUAIL for instance, is the QAI laboratory that supports NASA in its space missions.

Meanwhile quantum methods for machine learning expand (i.e. with TensorFlowQ, NetKet, Quantum Enhanced Machine Learning-QEML and other systems), the spatial contexts of quantum-enhanced machine learning become all the more important and spatially-sensitive (topological) methods of quantum computing have already been adopted to build machine learning algorithms (Aaronson et al., 2016). But, despite the interest in quantum approaches to human decision making (Busemeyer & Bruza, 2012; Busemeyer et al., 2006), spatial decision making is still short of equivalent QAI applications. Further, QAI requires algorithms to be created anew for every task and its accuracy decreases as the number of quantum bits increases. It also needs extremely low temperatures to support the hardware and (to this date) it is not completely error-free. Yet, it is anticipated that once it becomes widespread, all hitherto known cryptographic procedures may not suffice to prevent hacking based on quantum computing.

7.2 Big Data and Spatial Transcomputation

The term "geospatial data" is meant to include any data created from maps, field observations, photographs, national archives, satellite imagery (earth-observation and meteorological satellites), historical cartographic data, sensor networks, GNSS (GPS etc.), scanners (e.g. LIDAR), drones, or from combinations of such sources. The rapid expansion of geospatial technologies, particularly during the last decade, has brought about an unprecedented growth in big geospatial data, thus challenging the methods and technologies that are used to store, retrieve and process geospatial information. Some years ago, big geospatial data were estimated to increase by 20% annually (Lee & Kang, 2015). NASA alone collects quite a few petabytes of geospatial data and ESA is estimated to collect 10 terabytes daily, whereas a drone such as "MQ9 Reaper" yields 1, 4 gigabytes/sec of data (Jiang & Shekhar, 2017; Werner & Chiang, 2021). In response to the need to store, manage and retrieve information from geospatial big data, some specialized software appeared, beginning with Google's "MapReduce" in 2004, followed by Apache's "Hadoop" in 2006, the relational database "Postgres", "Geomesa, "Geowave", and IBM's "GeoscopePairs" which can handle petabytes of raster and vector geospatial data etc. Simultaneously, certain standards that are useful for processing geospatial big data have been established, such as CSV (for row-based data), JSON (for row-based, with Javascript Object Notation), AVRO (row-based), PARQUET and ORC (column-based). Besides these, Support Vector Machines (SVM) have also been used to classify spatial data from satellite imagery (Basheer et al., 2022) and it is interesting that SVMs may also be combined with swarm optimization algorithms (Soliman et al., 2012). Handling big data presents a number of challenges, such as compression, interoperability, standardization, discoverability, reliability, uncertainty, metadata management, licensing etc. (Sunitha & Sivarani, 2021). But the incessant growth in geospatial information has not been promulgated by the corporate or state sectors only. In fact, some large big geospatial datasets emerged as a result of the expansion of "neogeography" which enables anyone to create map "mashups", by the expansion of geotagging and "volunteered geographical information" (Papadimitriou, 2010a, 2010b). These were made possible by using websites such as "Panoramio", "Platial", "Flickr", "Sharing Places", "Zoomr" and, expectedly, the spontaneous uploading of volumes of geo-referenced photographs and texts will contribute to further increasing the total volume of spatial datasets in the future.

Furthermore, both "geospatial" and "spatial" big data (Barnes & Wilson, 2014; Yang et al., 2019) inevitably entail the fundamental problem of complexity (Robinson et al., 2017), which becomes particularly explicit in applied geographical analysis in the case of large landscape databases (Zhang & Li, 2022) and spatial analysis (Papadimitriou, 2020a, 2020b, 2020c).

Yet, "big spatial data" does not necessarily imply difficulty in data mining, since it does not always imply high spatial complexity as well. This is particularly important for geospatial big data, because it is precisely the overall "complexity" (algorithmic and computational) that eventually matters for information storage and processing. It is encouraging that if the location of a particular cell on a square map were to be identified by means of successive binary questions that partition the map into halves iteratively, then the information required tends to zero as the map size tends to infinity. This is the "miracle of spatial information" (Papadimitriou 2022, 2022a): as the size n of a square map increases, less and less spatial questions are needed to identify the location of particular square cell on it:

$$\lim_{n \to \infty} \frac{I}{n} = 0$$

So, in theory at least, there is hope that geospatial big data in raster format can be explored in sufficient detail. However, there may be other limits to spatial analysis.

In fact, there are physical limits to spatial computing and, by consequence, to any AI, which might also include QAI (Grabowska & Gunia, 2024). The Margolus-Levitin theorem for instance, asserts that no computational speed per unit of energy can exceed 6×10^{33} operations per Joule per second. And, it takes more than 1000 MWh of energy to train ChatGPT (Patterson et al., 2021), to the extent that concerns about the energy requirements of AI have even led to carbon footprint assessments for the function of big AI installations (Anthony et al. 2020).

The Landauer limit is the minimum amount of energy required to change one bit of information (Landauer, 1961; Bennett, 1973, 1982; 2003; Vaccaro and Bennett, 2011; Bérut et al., 2012). It is a direct consequence of the second law of thermodynamics (postulating that the entropy of a closed system cannot decrease) and is equal to:

$$E_L = k_B T \ln 2.$$

where k_B is the Boltzmann constant $k_B \sim 1.38 \times 10^{-23}$ Joule/deg.Kelvin, T is the absolute temperature in degrees Kelvin and $\ln 2 = 0.69315...$ So at room temperature 25 °C (298° Kelvin), the Landauer limit is 285×10^{-23} Joule. This is the minimum amount of energy coresponding to flipping 1 bit of information: if one bit is lost, the amount of energy emitted to the environment is $E \geq k_B T \ln 2$ which is the energy dissipation for each (irreversible) bit operation. Alternatively, this limit is equivalent to the energy cost of erasing 1 bit of infrormation and if n bits of information are erased, then the energy is at least $n k_B T \ln 2$ Joule. Theoretically, the temperature in which the computation takes place can not be made lower than 3 deg.Kelvin (equal to the approximate temperature of the cosmic microwave background radiation), because more energy should be spent on cooling than what would be saved in computation. Although concerns have been expressed as to whether ASI will ever be realized with classic computation technologies, precisely due to the very high energy requirements (Stieffel & Coggan, 2023), the Landauer limit may not apply to reversible computing and, from the estimations that have hitherto been published (Esposito, 2018; Neto & Bernardo, 2024; Wang, 2022), it is still not completely clarified whether it applies to quantum computation either.

Aside of energy however, assessing the computational power required to compute the total number of possible binary square map configurations of size n, symbolized as $N(n)$, for varying map sizes can be quite challenging and may also indicate limits to spatial computation. Certainly, the number of operations that AI technologies may perform is not expected to be the same all accross the range of AI types and technologies (and it isnt), since it depends on the particular parameters that each AI technology uses. In the case of CNNs for instance, the number of operations may not depend directly on the number of parameters (Desislavov et al., 2023). Characteristically, an assessment of the computational limits of AI in this respect (Thompson et al., 2020) revealed that a tenfold increase in performance requires 10,000 more computations.

The world's currently more powerful computer (at the time of writing this book), the "Frontier" at the Oak Ridge Laboratory, can perform approximately 1.102×10^{18} (1102 quintillion) operations per second. A "gedankenexperiment" with the simplifying assumption that the calculation of one binary map configuration can be made as *one* computer operation only, then for 3×3 maps, the time required to compute all 3×3 binary map configurations is approximately $2^9/(1.102 \times 10^{18})$ sec, that is 4.6×10^{-16} sec, while for all 4×4 maps it would be 5.95×10^{-14} sec. These time scales may appear deceivingly small when map sizes are small. Because, calculating the time required for *all* possible square binary map configurations of the earth's surface at a resolution of 1km x 1km (by classical computing) may be surprisingly large. Since there are

$$N(n) = 10^{10^{8.82\cdots}}$$

possible binary map configurations of the earth's surface at a resolution 1kmx1km (Papadimitriou, 2020a, 2020b), each such raster map would have $n = 5.1 \times 10^8$ square cells of area 1 km², and hence it follows that this calculation would require $9.23452 \times 10^{660693429}$ sec. As the age of the universe is estimated to be (approximately) 4.415×10^{17} sec, it follows that the calculation time (with the previously stated simplifying assumption and by using the same supercomputer) for all possible alternative square binary map configurations as large as the earth' surface would require many times the age of the universe.

Similarly, the maximum number of operations that can be computed by "Frontier" in time that is less than or equal to the age of the universe is

$$2^{mapsize} = \frac{(1.102 \times 10^{18} \text{operations})}{s}(4.415 \times 10^{17}\text{s})$$
$$= 4.865 \times 10^{35}\text{ops}$$

and thus the maximum number of binary square map configurations would be $2^{118.55}$ which implies that the maximum map size may only be as small as 118×118.

Besides energy and time however, the Bremermann limit (Bremermann, 1967) sets an upper bound to the total number of bits that can be processed by a hypothetical computer that has the size and age of our planet: the limit is 10^{93} bits so any problem that leads to higher figures of information bits is called "transcomputational". As it has turned out in multiple instances in the physics of computation, this limit can be easily attainable if the problem at hand involves combinatorial explosions.

Adopting the equivalent of a "brute force" approach (in spatial terms) for the identification of the location of each cell on a raster map requires (Papadimitriou, 2022a):

$$I_0 = \begin{bmatrix} 2 + 2\lfloor \log_2 \sqrt{n} \rfloor & \text{iff } \log_2 \sqrt{n} \notin Z \\ 2\log_2 \sqrt{n} & \text{iff } \log_2 \sqrt{n} \in Z \end{bmatrix}$$

bits, so if it is a square map, the information is

$$I_0 = 2\log_2 \sqrt{n} \text{ bits.}$$

where n is the map size. This formula is derived from the number of binary questions required to identify the position of any cell in a raster map. Each question asks

Fig. 7.1 For a chessboard-type map (8 × 8), it takes 6 successive binary questions in order to identify the location of one cell in it (questions numbered from 1 to 6)

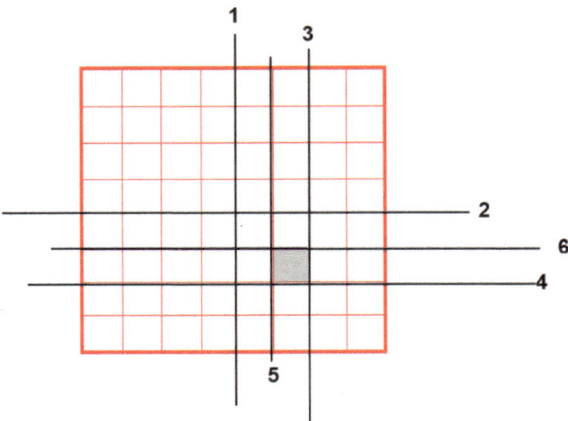

whether the cell is on the left or right half of the map (e.g. the 1st question), whether it is in the upper or at the lower part of the map (2nd question) etc. For instance, for a chessboard ($n = 64$) are required 6 questions, so the information corresponding to the identification of the location of a cell is tantamount to $I_0 = 6$ bits (Fig. 7.1).

Considering that there are 2^n possible binary square map configurations of size n, the total amount of information required to identify each and all cells in all possible configurations of square binary maps of that size (n) is

$$I_n = 2^n I_0 = 2^{n+1} \log_2 \sqrt{n} \ \text{ bits.}$$

Observing the Bremermann limit of 10^{93} bits, the total information should therefore be

$$I_n = 2^{n+1} \log_2 \sqrt{n} < 10^{93} \ \text{ bits.}$$

Solving for n yields $n = 305.894$, and thus the side of the square map should have a length of

$$\lfloor \sqrt{n} \rfloor \cong \lfloor 17.49 \rfloor = 17 \ \text{ cells.}$$

Indeed,

$$I_n = 2^{289} \log_2 \sqrt{289} = 8.13 \times 10^{87} < 10^{93} \ \text{ bits.}$$

So any spatial optimization that purports to locate each and all individual cells in *all* possible binary configurations of such maps, should not exceed the size of 17 × 17 for classical computation.

Following a different rationale of assessment (Kureichik et al., 2020), it was suggested that the search becomes transcomputational for binary maps of size 18 ×

18 and larger, while if there are more than two colors, transcomputation emerges at even smaller map sizes.

By extension to maps with x colors, and since there are x^n map configurations of square maps with size n, the total information required is

$$I_n = x^n I_0 = 2x^n \log_2 \sqrt{n} \ \text{ bits.}$$

Thus, for instance, only up to $x = 8$ colors can be used for the sum total of all multicolored map configurations of size $n = 100$ (that is 10 x 10).

Presently, it appears questionable (if not unlikely) that an ASI might possibly be created that would be capable of exceeding the Bremermann limit if the spatial problem has to be solved as a *single* task. Otherwise put, if a spatial problem *can* be broken down to simpler ones of lower complexity each (and in a way that none of them will be transcomputational itself), then it *might* be possible to bypass the Bremermann limit, thus increasing chances for the realization of ASI.

References

Aaronson, S., Ben-David, S., & Kothari, R. (2016). Separations in query complexity using cheat sheets. In *Proceedings of the forty-eighth annual ACM symposium on Theory of Computing* (pp. 863–876).

Abdelgaber, N., & Nikolopoulos, C. (2020). Overview on quantum computing and its applications in artificial intelligence. In *2020 IEEE third international conference on artificial intelligence and knowledge engineering (AIKE)* (pp. 198–199). IEEE.

Anthony, L. F. W., Kanding, B., & Selvan, R. (2020). Carbontracker: Tracking and predicting the carbon footprint of training deep learning models. arXiv: 2007.03051.

Ayoade, O., Rivas, P., & Orduz, J. (2022). Artificial intelligence computing at the quantum level. *Data, 7*(3), 28.

Ayoade, O., Rivas, P., Orduz, J., & Rafi, N. (2023). Satellite image classification using quantum machine learning. In *Artificial intelligence in earth science* (pp. 337–355). Elsevier.

Barnes, T., & Wilson, M. W. (2014). Big data, social physics, and spatial analysis: The early years. *Big Data & Society 1*(1).

Basheer, S., Wang, X., Farooque, A. A., Nawaz, R. A., Liu, K., Adekanmbi, T., & Liu, S. (2022). Comparison of land use land cover classifiers using different satellite imagery and machine learning techniques. *Remote Sensing, 14*(19), 4978.

Bennett, C. H. (1973). Logical reversibility of computation. *IBM Journal of Research and Development, 17*(6), 525–532.

Bennett, C. H. (1982). The thermodynamics of computation-a review. *International Journal of Theoretical Physics, 21*(12), 905–940.

Bennett, C. H. (2003). Notes on Landauer's principle, reversible computation and Maxwell's Demon. *Studies in History and Philosophy of Modern Physics, 34*, 501–510.

Bérut, A., Arakelyan, A., Petrosyan, A., Ciliberto, S., Dillenschneider, R., & Lutz, E. (2012). Experimental verification of Landauer's principle linking information and thermodynamics. *Nature, 483*, 187–189.

Bremermann, H. J. (1967). Quantum noise and information. *Proceedings of the Fifth Berkeley Symposium on Mathematical Statistics and Probability, 4*, 15–20.

Busemeyer, J. R., & Bruza, P. D. (2012). *Quantum models of cognition and decision.* Cambridge University Press.

Busemeyer, J. R., Wang, Z., & Townsend, J. T. (2006). Quantum dynamics of human decision-making. *Journal of Mathematical Psychology, 50*(3), 220–241.

Desislavov, R., Martínez-Plumed, F., & Hernández-Orallo, J. (2023). Trends in AI inference energy consumption: Beyond the performance-vs-parameter laws of deep learning. *Sustainable Computing: Informatics and Systems, 38*, 100857.

Esposito, M. (2018). Landauer principle stands up to quantum test. *Physics, 11*(101103), 49.

Garg, S., Ramakrishnan, G. (2020). Advances in quantum deep learning: An overview. arXiv:2005. 04316.

Ghosh, A., Saha, S., Bhagat, V., & Ganie, A. H. (2024). Quantum vs. Classical: a rigorous comparative study on neural networks for advanced satellite image classification. In *2024 International conference on trends in quantum computing and emerging business technologies* (pp. 1–5). IEEE.

Grabowska, A., & Gunia, A. (2024). On quantum computing for artificial superintelligence. *European Journal for Philosophy of Science, 14*(2), 22–30.

Guo, M., & Wang, S. (2020). Quantum computing for solving spatial optimization problems. *High performance computing for geospatial applications* (pp. 97–113). Springer.

Henderson, M., Shakya, S., Pradhan, S., & Cook, T. (2020). Quanvolutional neural networks: Powering image recognition with quantum circuits. *Quantum Machine Intelligence, 2*(1), 2.

Jiang, Z., & Shekhar, S. (2017). *Spatial Big Data Science.* Springer.

Kureichik, V. V., Kursitys, I. O., Kuliev, E. V., & Gerasimenko, E. M. (2020). Application of bioinspired algorithms for solving transcomputational tasks. *Journal of Physics: Conference Series, 1703*(1), 012021.

Landauer, R. (1961). Irreversibility and heat generation in the computing process. *IBM Journal of Research and Development, 5*, 183–191.

Lee, J.-G., & Kang, M. (2015). Geospatial big data: Challenges and opportunities. *Big Data Research, 2*(2), 74–81.

Moret-Bonillo, V. (2015). Can artificial intelligence benefit from quantum computing? *Progress in Artificial Intelligence, 3*, 89–105.

NASA-QUAIL: Quantum artificial intelligence laboratory. https://ti.arc.nasa.gov/tech/dash/groups/quail/

Neto, C. O. A. R., & Bernardo, B. D. L. (2024). Efficient erasure of quantum information beyond Landauer's limit. ArXiv:2402.15812

Papadimitriou, F. (2010a). A "Neogeographical Education"? The geospatial web, GIS and digital art in adult education. *International Research in Geographical and Environmental Education, 19*(1), 71–74.

Papadimitriou, F. (2010b). Introduction to the complex geospatial web in geographical education. *International Research in Geographical and Environmental Education, 19*(1), 53–56.

Papadimitriou, F. (2022a). *Spatial entropy and landscape analysis.* Springer VS.

Papadimitriou, F. (2022b). Spatial entropy, geo-information and spatial surprise. *Spatial entropy and landscape analysis* (pp. 1–14). Springer VS.

Papadimitriou, F. (2020a). *Spatial complexity. Theory, mathematical methods and applications.* Springer.

Papadimitriou, F. (2020b). Geophilosophy and epistemology of spatial complexity. *Spatial complexity: theory, mathematical methods and applications* (pp. 263–278). Springer.

Papadimitriou, F. (2020c). Entering the "Spatium Numerorum": Creating spatial complexity from numbers. *Spatial complexity: theory, mathematical methods and applications* (pp. 143–161). Springer.

Patterson, D., Gonzalez, J., Le, Q., Liang, C., Munguia, L. M., Rothchild, D., So, D., Texier, M., & Dean, J. (2021). Carbon emissions and large neural network training. ArXiv: 2104.10350

Rivas, P., Zhao, L., Orduz, J. (2021). Hybrid Quantum Variational Autoencoders for Representation Learning. In *Proceedings of the 19th International Conference on Scientific Computing (CSC 2021)*, Las Vegas, NV, USA, 15–17 December 2021.

Robinson, A. C., Demšar, U., Moore, A. B., Buckley, A., Jiang, B., Field, K., Kraak, M. J., Camboim, S. P. & Sluter, C. R. (2017). Geospatial big data and cartography: research challenges and opportunities for making maps that matter. *International Journal of Cartography, 3*(sup1), 32–60.

Shokry, A., & Youssef, M. (2021). Towards quantum computing for location tracking and spatial systems. In *Proceedings of the 29th international conference on advances in geographic information systems* (pp. 278–281).

Soliman, O. S., Mahmoud, A. S., & Hassan, S. M. (2012). Remote sensing satellite images classification using support vector machine and particle swarm optimization. In *2012 Third international conference on innovations in bio-inspired computing and applications* (pp. 280–285). IEEE.

Stieffel, K. M., & Coggan, J. S. (2023). The energy challenges of artificial superintelligence. *Frontiers in Artificial Intelligence, 6,* 1240653.

Sunitha, T., & Sivarani, T. S. (2021). An efficient content-based satellite image retrieval system for big data utilizing threshold based checking method. *Earth Science Informatics, 14*(4), 1847–1859.

Thompson, N. C., Greenewald, K., Lee, K., & Manso, G. F. (2020). The computational limits of deep learning. *arXiv preprint* arXiv:2007.05558, *10.*

Vaccaro, J., & Barnett, S. (2011). Information erasure without an energy cost. *Proc. Royal Society A, 467*(2130), 1770–1778.

Wang, F. Z. (2022). Breaking Landauer's bound in a spin-encoded quantum computer. *Quantum Information Processing, 21*(11), 378.

Werner, M., & Chiang, Y. Y. (Eds.). (2021). *Handbook of big geospatial data.* Springer.

Yang, C., Yu, M., Li, Y., Hu, F., Jiang, Y., Liu, Q., Sha, D., Xu, M., & Gu, J. (2019). Big earth data analytics: A survey. *Big Earth Data, 3*(2), 83–107.

Zhang, C., & Li, X. (2022). Land use and land cover mapping in the era of big data. *Land, 11*(10), 1692.

Zhu, Y., & Yu, K. (2023). Artificial intelligence (AI) for quantum and quantum for AI. *Optical and Quantum Electronics, 55*(8), 697.

Chapter 8
AI and Spatial Complexity

Abstract AI can be used to explore fundamental properties of spatial complexity that are characteristic of some particular map size and entropy class irrespective of the population they represent, i.e. to prove that the values of spatial complexity increases in 4×4 binary maps as their spatial entropy increases, attains a maximum and then decreases. Yet, it is uncertain how might AI cope with intractable spatial problems which may emerge from land use optimisation or from spatial (board) games. In fact, it is shown that spatial complexity may affect computability in cases of even small domains. Nevertheless, new mathematical methods may enable us to tackle computationally hard problems that might hamper the growth of Spatial AI in the future.

Keywords Spatial AI · Spatial complexity · Spatial computability · Computability · Evolutionary AI · 4×4 · Binary maps · Square maps

8.1 Spatial Complexity Assessment with AI

As much as identifying features and patterns in images is (and remains) an important facet of AI applications in spatial analysis, discovering fundamental properties of spatial allocations is another. Essentially, a central question in handling large geospatial/geographical/spatial datasets is to explore the extent to which some of their essential spatial topological and geometric properties can be inferred from small such datasets. This question essentially reflects another more fundamental issue that is deciding how spatial characteristics of small maps can be indicative of the properties of larger ones. Next, it will be shown how a software that has been developed on the basis of genetic algorithms can be useful in exploring this question. Consider the following problem of spatial analysis: What is the average spatial complexity and aggregate size of 4×4 binary square maps? Confining this problem to square binary maps of that size, and in order to find numerical relationships among spatial complexity, spatial entropy and aggregate size, a large number of calculations needs to be carried out on such maps; to get a glimpse of this magnitude, it suffices to

F. Papadimitriou, *Spatial Artificial Intelligence*,
SpringerBriefs in Computational Intelligence,
https://doi.org/10.1007/978-3-031-82136-3_8

consider the number of possible binary map configurations (Papadimitriou, 2020a, 2020b, 2020c, 2022) up to maximum spatial entropy class (which is $r = n/2 = 8$ in this case):

$$\sum_{r=1}^{r=n/2} \binom{n}{1} = \binom{16}{1} + \binom{16}{2} + \cdots + \binom{16}{8}$$
$$= 16 + 120 + 564 + \cdots + 13520 = 36.493$$

(the "entropy class" of a binary map is the number of black cells in the map). At most a quarter of these configurations are topologically equivalent to one another and therefore yield the same results, because the application of square symmetry operations (rotations, translations, inversions) produces equivalent positions of cells, since there are combinations of cells that yield topologically equivalent maps (Papadimitriou, 2020a, 2020b, 2020c). Consequently, all calculations need to be carried out on non-isomorphic maps for each spatial entropy class, which are taken by random sampling, i.e. with at least 10% per entropy class (Table 8.1).

Hence, a set of 289 maps was analyzed. The spatial complexity measure C_{P2} was calculated for each and all of the selected map configurations (for the procedure of calculation, the reader is referred to Papadimitriou, 2020a, 2020b). At this point, the software "Eureqa"™ is introduced that implements evolutionary computation on the basis of genetic algorithms. It was created at Cornell Creative Machines Lab (Schmidt & Lipson, 2009), has been commercialized by Nutonian Inc. and a similar version is currently available ("Turingbot"). In order to approximate the functions that best describe the observational data, the software yields alternative equations fitting the data it is fed with. It is an AI software for symbolic regression (Dubčáková, 2011; La Malfa et al., 2021; Lin, 2009; Stoutemyer, 2013) that has the capacity to identify alternative equations that model the relationships among variables and parameters and associates each equation with a complexity value and an error estimate. Consequently, the user may select i.e. an equation with high accuracy albeit

Table 8.1 The sampling procedure: at least 10% of all non-isomorphic map configurations were sampled per spatial entropy class

Spatial Entropy class r	Possible map configurations	Of which non-isomorphic (at least 1/4 of the possible configurations)	Of which sampled (%)	Sample size
1	16	2	10	0
2	120	15	10	2
3	564	71	10	7
4	1820	228	10	23
5	4368	546	10	55
6	5005	626	10	63
7	11,080	1385	10	139

with high complexity, or an equation with low complexity but with low accuracy, or any other alternative equation in between (a decision has to be made by the user which function to select from the set of alternative models, in view of striking an optimum between minimal error and minimal function complexity). It is due to this unique type of modelling that this AI software has been applied in multiple instances (Al-Subhi, 2020, 2023a, 2023b; Apua, 2023; Ashok et al., 2021; Lee et al., 2020; Richmond-Navarro et al., 2022; Wasik et al., 2015). Hence, a solution can be selected (Table 8.2, Fig. 8.1) to model the number of possible combinations of cells N_r as they increase with increasing spatial entropy class:

$$N_r = \exp(0.381r)$$

Similarly, the relationship between the number of aggregates (or possible blocks) N_A per aggregate size (A) can be approximated by a cubic function of A:

$$N_A = 0.187A^3 - 2A + 6.93$$

From this relationship, it can be deduced that the larger the aggregate, the higher the number of possible combinations of that aggregate size. As larger aggregates also imply higher spatial entropy (because more black cells occupy the map), this finding can be re-stated as: the higher the spatial entropy class, the higher the number of possible aggregates. Equivalently, the higher the number of aggregates, the higher the number of their possible spatial allocations. But a higher number of aggregates does not necessarily imply that the spatial complexity of such maps should increase as well. This point can be examined in more detail by measuring the spatial complexity (using the C_{P2} measure) of all the maps of the sample dataset. Consequently, approximating the change of frequency of C_{P2} values (F_{CP2}) with increasing values of C_{P2} yields:

$$F_{CP2} = 1.95C_{P2}\cos(0.46 + 0.73C_{P2}) + 3.68C_{P2} - 0.79C_{P2}\cos C_{P2} - 3.96$$

Table 8.2 Equations with increasing complexity and various error values for the data relating N_t and spatial entropy class r

$N_n = f(r)$	Error	Complexity
$N_n = 1.83r$	0.402	3
$N_n = \exp(0.381r)$	0.071	5
$N_n = \exp(0.381r) - 0.144$	0.062	7
$N_n = 1.54\exp(0.381r) - 0.144$	0.053	9
$N_r = 1.12\exp(0.317 + 0.333r) - 1.15$	0.053	11
$N_r = 1.17\exp(0.366r) - 0.695 - \frac{0.0326}{0.00734r^2 - 0.292}$	0.021	20

Fig. 8.1 a The number of possible combinations of cells N_r in 4 × 4 maps increases with their spatial entropy class, that is with the number (n) of their black cells. **b** The number of possible aggregates N_a in 4 × 4 maps increases with aggregate size A. **c** The frequency of C_{P2} values (F_{P2}) against C_{P2} values displays a maximum at $C_{P2} = 8$

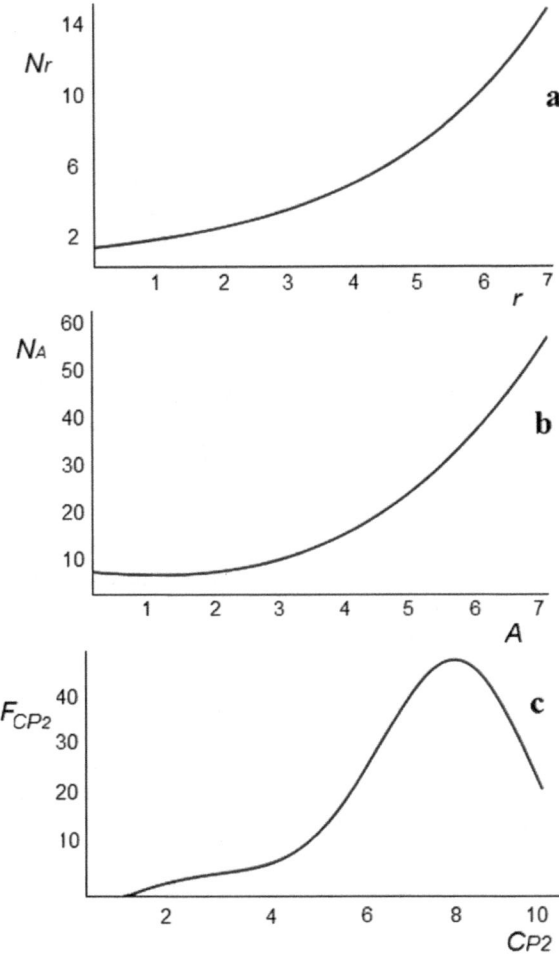

This function thus reveals that of all spatial entropy classes of varying aggregate sizes on 4 × 4 binary maps, the more frequently occurring spatial complexity values are those around $C_{P2} = 8$; a value which (not by accident) coincides with $C_{P2max}/2$.

In this way, using evolutionary computing it has become possible to infer some fundamental spatial properties of binary maps from small samples and hence, aided by an Evolutionary AI software, it has become possible to unveil and explore fundamental characteristics of spatial complexity; "fundamental" because they characterize some particular map size and number of colors *irrespective* of the population they represent. Evidently, expanding this research by sampling even larger numbers of map configurations, multicolored maps and larger map sizes (5×5 etc.) may corroborate further the usefulness of Evolutionary AI in discovering basic properties of spatial allocations.

8.2 AI and Computational Complexity

Aside of restrictions that are possibly imposed by the physics of information, there may exist mathematical limitations to Spatial AI as well and these relate to solvability and computability. As the history of scientific ideas has shown, the road from spatial simplicity to spatial complexity has led to the realization that some problems are non-computable.

Spatial computability problems can be partially tackled, by taking into account some parameters at the expanse of other ones. Mathematical formalism is a sine qua non towards this aim, but, simultaneously, it can be overly perplexing also. A characteristic example for spatial objects created in some virtual space is the "rendering equation" (no spatial entity can be placed in a well-structured VR or AR space without deciding its viewing point, angle and illumination). The "rendering equation" is the key mathematical formula for rendering (Kajiya, 1984, 1986, 1989; Immel et al., 1986): a master equation giving the radiance emitted from a point x as the sum of the emitted and reflected radiances. At each position x and direction w, the outgoing light L_0 is the sum of the emitted light L_e and the reflected light L_r. The latter is the sum of the incoming light from all directions L_i, multiplied by the surface reflection and the cosine of the incident angle w' for the wavelength k and the hemisphere Ω:

$$L_0(x, w, k, t) = L_e(x, v, k, t) + \int_\Omega f(x, w, w', k, t) L_i(x, w'k, t)(-w'n)dw'$$

where f is the "bi-directional reflectance distribution function", which is the proportion of light reflected from w' to w at position x, at time t and at wavelength k and $-w'n$ is the attenuation of the incoming light due to the incident angle w'. This equation is linear and rotation invariant and presupposes energy equilibrium, meaning that the energy input and output are equilibrated for the whole range of the hemisphere Ω. Yet, the problem with this equation is that it is unsolvable.

Unsolvability and intractability are two of the most puzzling oddities in both computer science and mathematics: two extreme conditions testing the upper limits of human and artificial intelligence. Spatial problems that are definitely unsolvable are those for which there exists a theorem that proves the impossibility to solve them. For those, it has been *proven* that they can not be solved (we possess undisputable proofs that they can not be solved), i.e. for Dehn's hypothesis that there is no algorithm that can possibly compute whether two "words" (sequences of symbols) correspond to the same element of an algebraic group.

In geographical analysis, intractability may easily emerge by considering the sum total of map configurations of a given size. The number of different possible allocations of colours (equivalent to land use/cover categories) to a raster map's cells has received considerable attention as a land use optimisation problem (Cao et al., 2011, 2012; Ding et al., 2021; Liu et al., 2013, 2015, 2022; Ma et al., 2011; Masoomi et al., 2013; Rahman & Szabó, 2021; Seppelt & Voinov, 2002; Wang et al., 2021; Yang et al., 2012; Yuan & Liu, 2014; Zhang et al., 2016). In these studies, several different methods (mathematical, algorithmic, computational, including methods of Evolutionary AI) have been devised to tackle the problem of how to optimally allocate different land uses to different cells, considering the restrictions (geological, environmental, economic etc.) that may apply to each different cell. If n is the number of land uses and N the number of map cells, then, for large N, the calculation of the number of possible map configurations becomes intractable.

But beyond intractability due to combinatorial explosions, even small spatial domains can present surprising difficulties in computability, if they have high spatial complexity. In fact, this is very easy to verify. Consider a 5 × 5 map, with two or three colors and with regularities in the spatial allocations of these colours; in this case, it is easy for both humans and AI to predict what the "next" colours will be if all the map's cells were extended to the east and south (Fig. 8.2). Yet, if the colour allocation is random and there is no regularity at all, then, it will be completely impossible for either AI or a human intelligence to figure out what colour will each cell of the extended domains have.

Besides geographical analysis, several spatial games are known to generate lots of possible combinations. The number of states of the traditional game Go is about 10^{170} and for StarCraft (by Blizzard Entertainment, 1998) it is more than 100^{685} (Usinier et al., 2016). To get a glimpse of these magnitudes, it suffices to observe how much higher they are than the estimated number of protons in the observable universe is (10^{80}). However, these magnitudes did not prevent the AI machine AlphaGo from prevailing over the world's Go champions twice (in 2016 and 2017), echoing Gary Kasparov's defeat in chess by IBM's DeepBlue ten years earlier.

Such spatial games belong to different computational complexity classes and an advanced Spatial AI will confront different levels of difficulty in playing each one of these games. The class *PSPACE* is the set of problems that can be solved in polynomial space and is equivalent to its nondeterministic variant, the *NPSPACE* (Savitch, 1970). *EXPSPACE* is the set of problems that are solvable in $O(2^{p(n)})$ space, where $p(n)$ is a polynomial function of n and many spatial games belong to the *EXPSPACE* complexity class (including chess). Yet, despite geometric complexity theory striving to prove that $P \neq NP$, it is known that some problems involving raster image analysis can be *NP*-hard (Coeurjolly et al., 2008; Sivignon & Coeurjolly, 2009), while some spatial games are *NP*-hard or *NP*-complete as well, such as Minesweeper (Kaye, 2000). In fact, two of the most widely known spatial problems, the Traveling Salesman Problem and the Map Coloring Problem, are both *NP*-complete and other spatial optimization problems are *NP*-hard (e.g. assigning n facilities to m locations by minimizing total weighted assignment cost).

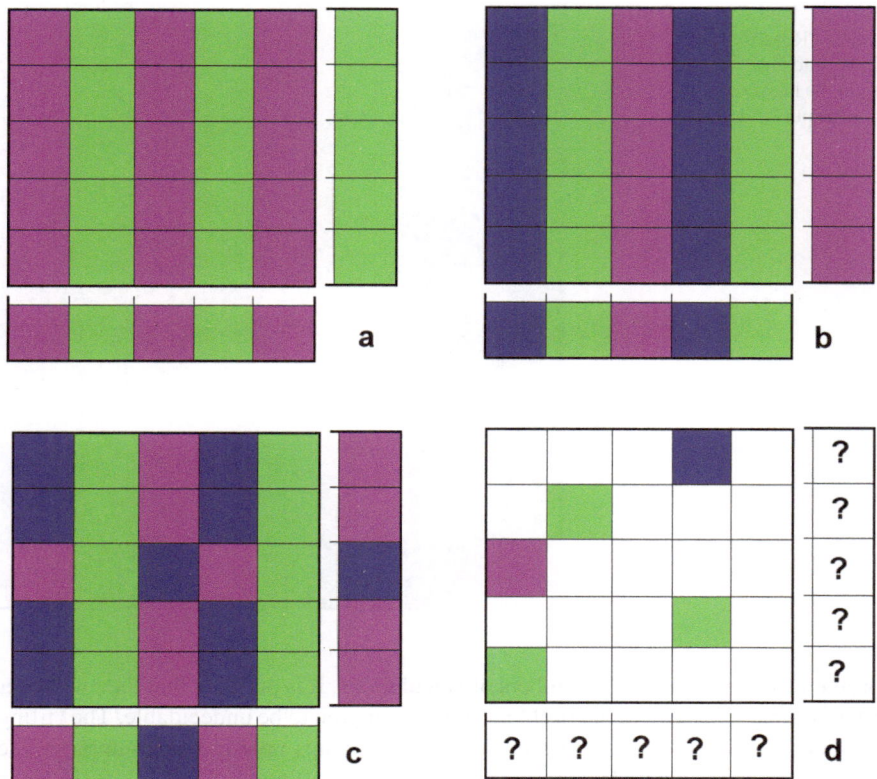

Fig. 8.2 Spatial complexity and spatial computability: if a map is entirely regular and symmetric (cases **a** and **b**) then it is fairly easy for AI to predict what the map's colors will be if its spatial domain is extended e.g. to the east and south. If the map is less regular (**c**), then it may not be easy to predict, but nevertheless feasible (e.g. by observing regularities of color changes). But if the map has high spatial complexity and the color allocations are completely random (**d**), it is impossible to predict the colors over the extended spatial domains

How will AI cope with spatial problems that are *NP-*(complete or hard) or intractable by humans will weigh heavily on the future developments of Spatial AI towards AGI and ASI. Whether an AI would be able to solve them or not opens up a wide field of research that, no doubt, will entail miraculous developments and surprises all along the way (Fig. 8.3). Meanwhile, the supposed potential of QAI for solving spatial *NP*-problems remains to be verified in practice (Otgonbaatar & Kranzlmüller, 2024).

By all calculations however, handling spatial information will become all the more efficient in the years and decades to come, so that a spatially-enabled AGI might eventually appear in the future, at a time by which we may expect machines to have successfully passed a "Turing test" of spatial intelligence, a "spatial Turing test". However, even if QAI or AGI eventually succeed to provide solutions to spatial

problems that require large numbers of calculations, it is unlikely that they will ever become able to tackle spatial problems that are proven to be undecidable. The "tiling problem" and the "worn stones problem" are perhaps appropriate examples. The tiling problem is an undecidable problem (Berger, 1966; Robinson, 1971) since there is no algorithm that solves it in all cases. For instance, although we know that equilateral triangles, squares and regular hexagons can tile the plane without leaving any "hole", determining whether a given finite set of tiles admits at least one valid tiling is unknown. The problem of the shape of worn stones is to decide what is the first shape that a stone takes if it is subjected to abrasion or erosion (i.e. does it take the form of a sphere or of an ellipsoid or some other shape?). This millenia-old problem, possibly attributable to Aristotle (Krynine, 1960), was put in rigorous form in the 20th c. (Firey, 1974) and is still open. None of these problems belongs to any known computational complexity class and they are still open.

How to circumnavigate physical limits to spatial computation? Simply put, by discovering mathematical methods that will downsize calculations in terms of space and time complexity. Among the several possible options, the following two appear more likely: i) reducing the number of operations required: e.g. despite the fact that the multiplication of two $n \times n$ matrices requires $2n^3$ additions and multiplications, the number of operations can be reduced if Strassen's method is used, bringing it down to $n^{\log 7}$; ii) reducing dimensionality: e.g. a set in $(n + 1)$-dimensional space defined by polynomials and inequalities can be projected onto n-dimensional space, and the resulting set will still be definable in terms of polynomial identities and inequalitie (by virtue of the Tarski–Seidenberg theorem).

Indeed, downsizing spatial complexity will be a sine qua non for circumnavigating limits to computation that might be imposed to advanced Spatial AI. Consider that the values of a square map are given by a matrix. The Grothendieck inequality asserts that if $A_{i,j}$ is an nxn (real or complex) matrix with

$$\left| \sum_{i,j} A_{i,j} s_i t_j \right| \leq 1$$

for all (real or complex) numbers s_i, t_j with absolute value at most 1, then there exists a constant, "Grothendieck's constant" k_G, with the property that

$$\left| \sum_{i,j} A_{i,j} \langle S_i T_j \rangle \right| \leq k_G$$

holds for all vectors S_i, T_j in the unit ball $B(H)$ of a real or complex Hilbert space and k_G is the smallest constant that satisfies this inequality for all nxn matrices. Its exact value remains unknown. Yet, if/when it will be calculated with exactness, it will pave the road towards solving many spatial problems. For instance, it may lead to easier detection of social network communities from random graphs and (more ambitiously perhaps) also allow for establishing correspondences between non-calculable (at least NP-hard) problems to other ones that are solvable in polynomial time.

Eugene Wigner, the founder of Cybernetics, first suggested that the power of mathematics to explain the world is puzzling, if not inexplicable altogether (Wigner, 1960), for it is a tool of human thought with which, beginning with few precisely defined premises, we can develop the capacity to create an intellectual edifice that turned out to be particularly effective to explain the world. His remarkable observation has been recalled in various domains, from ecology (Ginzburg et al., 2007) to data science (Halevy et al., 2009) and from physiology (Gorban et al., 2019) to AI in which it was shown that performance increases logarithmically with increasing the size of pre-training datasets (Sun et al., 2017). So beginning with "small" initial spatial datasets or logical rules, it is possible to explore disproportinately big spatial complexities. This often disregarded possibility strengthens the hope that future societies will be able to develop Spatial AI up to the physical limits of computation.

References

Al-Subhi, A. (2020). Parameters estimation of photovoltaic cells using simple and efficient mathematical models. *Solar Energy, 209*, 245–257.

Al-Subhi, A. (2023a). Accurate mathematical modeling of electronic load harmonics using machine learning software. *Yanbu Journal of Engineering and Science, 20*(1), 1–10.

Al-Subhi, A. (2023b). Mathematical-based models for solution of the load flow problem. *International Journal of Modeling, Simulation, and Scientific Computing, 14*(05), 2350026.

Apua, M. C. (2023). Development of leaching predictive models for elements extraction from coal fly ash in sulphuric acid solution: Application of Eureqa Newtonian software. *Journal of Chemical Technology and Metallurgy, 58*(1), 187–199.

Ashok, D., Scott, J., Wetzel, S. J., Panju, M., & Ganesh, V. (2021). Logic guided genetic algorithms (student abstract). *Proceedings of the AAAI Conference on Artificial Intelligence, 35*(18), 15753–15754.

Berger, R. (1966). The undecidability of the domino problem. *Memoirs of the American Mathematical Society, 66*, MR36–49.

Cao, K., Batty, M., Huang, B., Liu, Y., Yu, L., & Chen, J. (2011). Spatial multi-objective land use optimization: Extensions to the non-dominated sorting genetic algorithm-II. *International Journal of Geographical Information Science, 25*(12), 1949–1969.

Cao, K., Huang, B., Wang, S., & Lin, H. (2012). Sustainable land use optimization using boundary-based fast genetic algorithm. *Computers, Environment and Urban Systems, 36*(3), 257–269.

Coeurjolly, D., Hulin, J., & Sivignon, I. (2008). Finding a minimum medial axis of a discrete shape is NP-hard. *Theoretical Computer Science, 406*(1–2), 72–79.

Ding, X., Zheng, M., & Zheng, X. (2021). The application of genetic algorithm in land use optimization research: A review. *Land, 10*(5), 526.

Dubčáková, R. (2011). Eureqa: software review. Faculty of Safety Engineering, VŠB, Technical University of Ostrava (pp. 1–5) https://core.ac.uk/download/pdf/8987202.pdf (Accessed 11 Oct.2024)

Firey, W. J. (1974). Shapes of worn stones. *Mathematika, 21*(1), 1–11.

Ginzburg, L. R., Jensen, C. X., & Yule, J. V. (2007). Aiming the "unreasonable effectiveness of mathematics" at ecological theory. *Ecological Modelling, 207*(2–4), 356–362.

Gorban, A. N., Makarov, V. A., & Tyukin, I. Y. (2019). The unreasonable effectiveness of small neural ensembles in high-dimensional brain. *Physics of Life Reviews, 29*, 55–88.

Halevy, A., Norvig, P., & Pereira, F. (2009). The unreasonable effectiveness of data. *IEEE Intelligent Systems, 24*(2), 8–12.

Immel, D. S., Cohen, M. F., & Greenberg, D. P. (1986). A radiosity method for non-diffuse environments. *Sisgraph, 1986*, 133.

Kajiya, J. (1986). The Rendering Equation. *Sisgraph, 1986*, 143–150.

Kajiya, J., & Kay, T. (1989). Rendering fur with three dimensional textures. *Sisgraph, 1989*, 271–280.

Kajiya, J., & Von Herzen, B.P. (1984). Ray tracing volume densities, *Sisgraph, 18*, 165.

Kaye, R. (2000). Minesweeper is NP-complete. *The Mathematical Intelligencer, 22*(2), 9–15.

Krynine, P. D. (1960). On the antiquity of "sedimentation" and hydrology (with some moral conclusions). *Geological Society of America Bulletin, 71*(11), 1721–1726.

La Malfa, G., La Malfa, E., Belavkin, R., Pardalos, P. M., & Nicosia, G. (2021). Distilling financial models by symbolic regression. *International Conference on Machine Learning, Optimization, and Data Science* (pp. 502–517). Springer.

Lee, A. C., Huang, R. Y., Nguyen, T. D., Cheng, C. W., & Tsai, M. C. (2020). Laser powder bed fusion of multilayer thin-walled structures based on data-driven model. *Journal of Laser Micro Nanoengineering, 15*(1), 1–7.

Ligmann-Zielinska, A., Church, R. L., & Jankowski, P. (2008). Spatial optimization as a generative technique for sustainable multiobjective land-use allocation. *International Journal of Geographical Information Science, 22*(6), 601–622.

Lin, E. (2009). Eureqa, the Robot Scientist. Available online : http://www.physorg.com/news17939 4947.html (Accessed 11 Oct 2024)

Liu, C., Deng, C., Li, Z., Liu, Y., & Wang, S. (2022). Optimization of spatial pattern of land use: Progress, frontiers, and prospects. *International Journal of Environmental Research and Public Health, 19*(10), 5805.

Liu, X., Ou, J., Li, X., & Ai, B. (2013). Combining system dynamics and hybrid particle swarm optimization for land use allocation. *Ecological Modelling, 257*, 11–24.

Liu, Y., Tang, W., He, J., Liu, Y., Ai, T., & Liu, D. (2015). A land-use spatial optimization model based on genetic optimization and game theory. *Computers, Environment and Urban Systems, 49*, 1–14.

Ma, S., He, J., Liu, F., & Yu, Y. (2011). Land-use spatial optimization based on PSO algorithm. *Geo-Spatial Information Science, 14*(1), 54–61.

Masoomi, Z., Mesgari, M. S., & Hamrah, M. (2013). Allocation of urban land uses by Multi-Objective Particle Swarm Optimization algorithm. *International Journal of Geographical Information Science, 27*(3), 542–566.

Otgonbaatar, S., & Kranzlmüller, D. (2024). Exploiting the quantum advantage for satellite image processing: review and assessment. *IEEE Transactions on Quantum Engineering, 5*, 3100309.

Papadimitriou, F. (2020a). *Spatial complexity: Theory, mathematical methods and applications.* Springer.

Papadimitriou, F. (2020a). The spatial complexity of 3 × 3 binary maps. *Spatial complexity: Theory, mathematical methods and applications.* (pp. 163–178). Springer.

Papadimitriou, F. (2020b). The algorithmic basis of spatial complexity. *Spatial complexity: Theory, mathematical methods and applications* (pp. 81–99). Springer.

Papadimitriou, F. (2022). *Spatial entropy and landscape analysis.* Springer VS.

Rahman, M. M., & Szabó, G. (2021). Multi-objective urban land use optimization using spatial data: A systematic review. *Sustainable Cities and Society, 74*, 103214.

Richmond-Navarro, G., Montenegro-Montero, M., Casanova-Treto, P., Hernández-Castro, F., & Monge-Fallas, J. (2022). Roughness sub-layer wind speed model for tropical wooded areas. *Wind Engineering, 46*(3), 759–766.

Robinson, R. M. (1971). Undecidability and nonperiodicity for tilings of the plane. *Inventiones Mathematicae, 12*, 177–209.

Savitch, W. J. (1970). Relationships between nondeterministic and deterministic type complexities. *Journal of Computation and Systems Science, 4*(2), 177–192.

Schmidt, M., & Lipson, H. (2009). Distilling free-form natural laws from experimental data. *Science, 324*(5923), 81–85.

Seppelt, R., & Voinov, A. (2002). Optimization methodology for land use patterns using spatially explicit landscape models. *Ecological Modelling, 151*(2–3), 125–142.

Sivignon, I., & Coeurjolly, D. (2009). Minimum decomposition of a digital surface into digital plane segments is NP-hard. *Discrete Applied Mathematics, 157*(3), 558–570.

Stoutemyer, D. R. (2013). Can the Eureqa symbolic regression program, computer algebra and numerical analysis help each other. *Notes of the American Mathematical Society, 60*, 713–724.

Sun, C., Shrivastava, A., Singh, S., & Gupta, A. (2017). Revisiting unreasonable effectiveness of data in deep learning era. In *Proceedings of the IEEE International Conference on Computer Vision* (pp. 843–852).

Usunier, N., Synnaeve, G., Lin, Z., & Chintala, S. (2016). Episodic exploration for deep deterministic policies : An application to StarCraft micromanagement tasks. arXiv:1609.02993.

Wang, W., Jiao, L., Jia, Q., Liu, J., Mao, W., Xu, Z., & Li, W. (2021). Land use optimization modelling with ecological priority perspective for large-scale spatial planning. *Sustainable Cities and Society, 65*, 102575.

Wasik, S., Fratczak, F., Krzyskow, J., & Wulnikowski, J. (2015). Inferring mathematical equations using crowdsourcing. *PLoS ONE, 10*(12), e0145557.

Wigner, E. (1960). The unreasonable effectiveness of mathematics in the natural sciences. *Communications in Pure and Applied Mathematics, 13*(I).

Yang, X., Zheng, X. Q., & Lv, L. N. (2012). A spatiotemporal model of land use change based on ant colony optimization, Markov chain and cellular automata. *Ecological Modelling, 233*, 11–19.

Yuan, M., & Liu, Y. (2014). Land use optimization allocation based on multi-agent genetic algorithm. *Transactions of the Chinese Society of Agricultural Engineering, 30*(1), 191–199.

Zhang, H., Zeng, Y., Jin, X., Shu, B., Zhou, Y., & Yang, X. (2016). Simulating multi-objective land use optimization allocation using Multi-agent system—A case study in Changsha, China. *Ecological Modelling, 320*, 334–347.